Control and Instrumentation Technology in HVAC:

PCs and Environmental Control

Control and Instrumentation Technology in HVAC:
PCs and Environmental Control

by
Michael F. Hordeski

Published by
THE FAIRMONT PRESS, INC.
700 Indian Trail
Lilburn, GA 30047

Library of Congress Cataloging-in-Publication Data

Hordeski, Michael F.
Control and instrumentation technology in HVAC: PCs and
environmental control / by Michael F. Hordeski.
    p. cm.
    Includes bibliographical references and index.
    ISBN 0-88173-306-7
1.   Heating--Automatic control. 2. Ventilation--Automatic control. 3. Air
conditioning--Automatic control. 4. Digital control systems. I. Title.
TH7466.5.H67   1999   697--dc21   99-049864
                                                                    CIP

*Control and instrumentation technology in HVAC: PCs and
environmental control / by Michael F. Hordeski.*

Published by The Fairmont Press, Inc.
700 Indian Trail
Lilburn, GA 30047

Printed in the United States of America

10 9 8 7 6 5 4 3 2 1

ISBN   0-88173-306-7   FP

ISBN   0-13-087995-9   PH

While every effort is made to provide dependable information, the publisher, authors, and
editors cannot be held responsible for any errors or omissions.

Distributed by Prentice Hall PTR
Prentice-Hall, Inc.
A Simon & Schuster Company
Upper Saddle River, NJ 07458

Prentice-Hall International (UK) Limited, London
Prentice-Hall of Australia Pty. Limited, Sydney
Prentice-Hall Canada Inc., Toronto
Prentice-Hall Hispanoamericana, S.A., Mexico
Prentice-Hall of India Private Limited, New Delhi
Prentice-Hall of Japan, Inc., Tokyo
Simon & Schuster Asia Pte. Ltd., Singapore
Editora Prentice-Hall do Brasil, Ltda., Rio de Janeiro

Preface

Building conditioning involves the use of a unit designed to keep occupied spaces comfortable or unoccupied spaces at desired levels of temperature and humidity. This unit generally consists of fans, heat-exchanger coils, dampers, ducts and the required instrumentation.

While most industrial controls have benefited from advances in personal computer control and sensor technology, building controls have lagged behind. Only now are some of the techniques used in industrial automation showing up in HVAC. The advantages are drastic reductions in operating costs due to the increased efficiency of operation.

In conventional systems, the amount of outdoor air admitted is usually excessive and based on criteria when energy conservation was given little attention. Some studies show that infiltration of outdoor air accounts for about 1/2 of the total heating and cooling loads. One study indicated that 75% of the fuel oil in New York City schools is used to heat ventilated air. Since building conditioning accounts for about 20% of the energy consumed in the U.S., optimized HVAC systems can make a major contribution in reducing our national energy use.

The normal operating modes include start-up, occupied, night and purge. Optimizing the start-up time involves automatically calculating the amount of heat that needs to be transferred. Computer-optimized controls can initiate the unoccupied mode of operation. The computer can be programmed for weekends and holidays. It also provides a flexible system for time-of-day control. This can be a low cost personal computer with the proper software.

A purge mode also allows a high degree of optimization. When the outside air is preferred to the inside air, the building can be purged. Free cooling is used on summer mornings and free heat is utilized on warm winter afternoons.

A computer controlled system also allows the heat storage of the building to be used for optimization. Purging during the night before a warm summer day allows the building structure to play a part in the optimization scheme.

Another optimization mode is summer/winter switching. A computer control system can recognize that summer-like days can occur in winter and cool days can occur in summer.

Optimized summer/winter operation takes place in an enthalpy-like basis. Heat is added in winter-like days and removed in summer-like days. A heat balance calculation can be made by the computer using the following parameters: cold- and hot-deck flows, exhaust airflow and enthalpy, cold- and hot-deck temperatures, mixed air temperature and outside air enthalpy.

Smart microprocessor-based thermostats are programmable with some memory capability. They can be monitored and reset by a central computer using wireless or cable communication links. The microprocessor-based units can have continuously recharged backup batteries and room temperature sensors that accurately monitor current conditions.

A major source of inefficient operation in conventional HVAC control systems is the uncoordinated operation of temperature control loops. There may be several temperature control loops in series. These can cause simultaneous heating and cooling operations to occur. A coordinated control system can reduce energy costs by 10% or more. Besides eliminating simultaneous heating and cooling, other interactions such as cycling are reduced or eliminated.

In many conventional HVAC systems, outdoor air is provided by keeping the outdoor air damper at a fixed open position when the building is occupied. In a computer control system the open position can be a function of the number of occupants of the building or the building's carbon dioxide levels.

A constant damper opening is incorrect since a constant damper opening does not provide a constant airflow. The flow varies with fan load. The conventional system wastes air conditioning energy at high loads and provides insufficient air at low loads. A computer control system can reduce operating costs with a constant minimum rate of airflow, which is unaffected by fan loading.

As energy costs continue to grow in relation to overall operating costs, the need for more refined HVAC control becomes more important. HVAC strategies such as optimizing start-up time and supply air temperature and minimizing fan energy and reheating are not only possible but are becoming necessary. This book examines the relationship between industrial automation techniques and evolving HVAC systems.

It should be of interest to HVAC system designers, specifiers, technicians and a wide variety of other readers.

The focus is on cutting-edge technologies and issues that can be applied to HVAC control. This includes evolving system and device buses, the Internet, intranets and enterprise links.

One control strategy sends plant alarms out over modem links. The plant engineer then logs in over the Internet to all nodes of the system. After checking an alarm summary to find the cause, corrections can be made to a simulated model of the system. The model then verifies the corrections before any actual changes are made to the system.

Another goal of this book is provide keys that open doors for additional information in the form of useful Internet resources and sites. The appendix includes a list of related Internet sources that can be used for additional details and a list of abbreviations found in the text.

Chapter 1 introduces heating and cooling systems. Topics include thermal comfort factors, personal factors, temperature and humidity control, dehumidification, intelligent HVAC, zoning, oil and gas furnaces, heat pumps, electric furnaces, combo units, hydronic heating systems and controls, water hammer, radiators, radiant floors, radiant electric environmental risks, air conditioning, geothermal heat pumps and solar systems.

Chapter 2 considers optimizing airhandler operations using techniques such as zone control, free cooling and summer/winter mode switching. Topics include motor starting, enthalpy and temperature measurement, fan control, fan energy, economizer cycles and optimizing strategies. VAV throttling and chimney effects are also considered.

Chapter 3 is concerned with boiler and pump control including boiler efficiency. This involves boiler instrumentation for operation monitoring and control requirements including two- and three-element control. Alarms and safety interlocks are considered along with boiler time response. Control techniques include feedforward control of steam pressure, fuel control, airflow control, damper and fan control, furnace draft control, fuel-air ratio control, feedwater control and pump speed control.

Chapter 4 considers heat pump and chiller optimization. This chapter starts with heat pump thermodynamics, refrigerants and refrigerator controls. Economizers are discussed along with load-following techniques. Cooling tower optimization techniques are considered as well as safety controls and flow balancing. Reconfiguration optimization techniques are included in this chapter along with free cooling techniques.

Environmental Controls are discussed in more detail in Chapter 5. This includes thermostatic controls, humidostats, and temperature sensors. The characteristics of analog controllers are explained including alarms and output limiting, reset windup, input signals, displays and control modes. Temperature controllers as well as position controllers are considered. Nonlinear controllers and balancing are discussed. Temperature regulators and thermal actuators are also covered in this chapter.

Wireless control has potential applications in HVAC and is explained in Chapter 6 starting with an introduction to light-beam control, photovoltaic cells, photoresistors and other photosensitive devices. Circuits for light-beam and infrared transmitters and receivers are described. Fiberoptic techniques are explained including fiberoptic networks and fiberoptic modems. Carrier-current control systems are a simple, inexpensive method of control. The characteristics of these systems are discussed as well as those of radio control and radio modems.

Computer control is introduced in Chapter 7 with the basics of programming and signal conversion. Personal computer and programmable logic controller applications and characteristics are covered in some detail. The concept of open systems is introduced with emerging standards such as LonWorks, BACnet, ControlNet and DeviceNet. Building automation trends are discussed along with open systems considerations.

Chapter 8 is concerned with building automation and distributed control techniques. Today's evolving controls depend on direct digital control in a distributed configuration for improved reliability. These systems depend on both high level buses for communication and low level busses for input and output.

The high level busses include TOP and the Manufacturing Automation Protocol (MAP). Low level buses include Profibus, FIP-BUS, MODBUS, Rackbus and Foundation Field bus for device access. Typical building automation software is discussed for such applications as optimized duty cycles, demand limiting, setback control, economizer cooling, and boiler and chiller optimization. Industrial intranets can provide the means of using web browsers for remote monitoring and diagnostics of HVAC systems.

Many thanks to Dee, who helped in getting both the text and the drawings in their final form.

Contents

Control, Reciprocating or Metering Pumps, Rotary Pumps, Safety and Throttling Control, Centrifugal Pumps, Pump Starters, On-Off Pressure Control, Throttling Control Valves, Pump Speed Throttling, Variable-speed Pumps, Variable-speed Drive Efficiency, Types of Variable-speed Drives, Multiple Pumps, Booster Pumps, Water Hammer, Optimizing Pump Controls, Valve Control Optimization, Blowdown, Load Allocation

Chapter 4 Heat Pump and Chiller Optimization ..103

Heat Pump Thermodynamics, Refrigerants, Refrigerator Control, On-Off Control, Commercial Chillers, Larger Chillers, Economizer, Safety Interlocks, Cooling Costs, Load-Following, Cooling Towers, Controls, Interlocks, Evaporative Condensers, Tower Optimization, Supply Temperature Optimization, Return Temperature Optimization, Safety Controls, Flow Balancing, Winter Operation, Evaporator Optimization, Water Supply, Temperature Optimization, Water Return Temperature Optimization, Tower Supply Temperature Optimization, Tower Return Temperature Optimization, Heat Recovery Optimization, Optimized Operating Mode, Reconfiguration Optimization, Free Cooling, Plate and Frame Heat Transfer, Direct Free Cooling, Mode Reconfiguration, Optimized Storage, Load Allocation, Feedforward Control, Retrofit Optimization, Economizer Control, Water Distribution

Chapter 5 Environmental Controls 133

Thermostatic Control, Two-position Control, Dual Thermostats, Setback Units, Heating-Cooling Thermostats, Limited Control, Zero Energy Band Control, Slave and Master Thermostats, Smart Thermostats, Pneumatic Units, Setpoints, Throttling Range, Thermostat Action, Sensitivity, Thermostat Design, Humidostat Design, Thermostat Accuracy, Temperature Sensors, Thermistors, Temperature Probes, Electronic Thermostats, Boiler Control, Temperature Equalization, Heater Humidifiers, Air Conditioner Humidity Controls, Controllers, Analog Controllers, Alarms and Output Limiting, Reset Windup, Input Signals, Setpoint Input, Displays, Control Modes, Electronic Temperature Controllers, Position Controllers, Nonlinear Controllers, Balancing, Servicing Features, Optical Meter Relays, Digital Controllers, Software, Temperature Regulators, Regulator Design, Thermal Actuators, Refrigerant Control

Chapter 6 Wireless Control 163

Light-beam Control, Photovoltaic Cells, Photoresistors, Other Photosensitive Devices, Light-controlled Relays, Light-beam Transmitter, Light-beam Receivers, Infrared Transmitters, Infrared Receivers, Multifunction Infrared Transmitters, Multifunction Infrared Receivers, Chopped Light Control, Light-controlled Capacitance, Fiberoptics, Fiberoptic Conductors, Fiberoptic Transmitters, Fiberoptic Receivers, Fiberoptic Connectors, Fiberoptic Cable, Fiberoptic Networks, Fiberoptic Modems, PLC Modems, PLC Optical Systems, Carrier-current Control Systems, Carrier-current Signal Frequency, Tone Encoding, Filtering, Carrier-current Transmitters, Carrier-current Receivers, Tone Decoders, Tone Encoding Control, Filter Decoding, PLL Decoding, PLL Operation, Noise and Response Characteristics, Touch-tone Encoding, Radio Control, Low-power AM Transmitters, Radio Modems, Ethernet Radio Modems, Wireless Networks, Cells, Standalone Cells, Ethernet Cells, Overlapping Cells, Roaming, Interference Considerations,

Table List

Figure List

Figure List (continued)

Chapter 1

Heating and Cooling Systems

Thermal Comfort Factors

There are several factors that influence thermal comfort. These are air temperature, air velocity, relative humidity, radiant environment and activity level. Air temperature is the most common measure of comfort, and the one that is most widely understood. When the air temperature rises or falls, it affects how much heat our bodies give up via conduction, convection, and evaporation.

Conduction is the transfer of heat through a single material or from one material to another where their surfaces touch. The rate at which heat is lost or gained depends on the nature of the materials and the difference in their temperatures. Your bare skin is at about 98.6°F. On a cold winter morning, the difference produces heavy conductive losses and you feel uncomfortable. A good wind will produce convective heat losses at the same time. This is typically called wind chill. The convective heat losses occur as cool air moves along the surface of your skin. Air movement inside a building can also affect comfort. Streams of air leaking through openings are known as infiltration. Indoor air movement is also created by natural convective currents (as warm air rises and cold air falls) and the drafts from forced-air heating systems, air conditioners, and fans. These interior winds can be a source of wintertime discomfort and in summer, most people would rather feel comfortable without the draft.

Air temperature has a major effect on how much heat we are able to shed by evaporation. Within certain temperature and humidity ranges, our sweat glands can produce over half a gallon of water per hour. As air passes over the skin, and that water evaporates, heat is drawn out of the body.

Both temperature and relative humidity work together in determining comfort. As the humidity rises, we are not as able to shed heat by sweating, so hot and dry conditions feel more comfortable than warm and humid weather.

Relative Humidity

Several comfort and health problems can occur from too much or too little humidity in the air. When the relative humidity indoors starts to exceed 50% on a continuing basis, molds, mildews, bacteria, and other bio-effects start to occur. Depending on the air temperatures outside and in, windows and walls may start to sweat.

A higher relative humidity usually helps in the winter because you lose less body heat through evaporation. A lower relative humidity is desirable in the summer, since you want all the evaporative cooling you can get. This means you can save on wintertime fuel by adding humidity to the indoor air, since you would feel comfortable at a lower setpoint on the thermostat. By dehumidifying indoor air in the summer, you can save on air conditioning and lose nothing in comfort.

If you exceed these balance points, too much or too little humidity can cause other problems. At relative humidities below 15%, throat and nasal passages can become dry, paint may crack and not hold to some surfaces, wood will shrink, and static electricity becomes a problem to sensitive electronics.

Radiant Temperature

The radiant temperature of your surroundings also affects how comfortable you feel. Most surrounding surfaces are cooler than the human body, and heat always radiates from warm surfaces to cold so we are usually losing radiant heat to the objects around us. This tends to help our comfort levels in the summertime, when we need to shed heat, but it works against us in the winter. The air temperature inside may be a very comfortable range in mid-winter, but if your body is losing enough radiant heat to a cold wall or to a large stretch of window, you are going to feel cool.

Insulated floors, walls, and ceilings keep their radiant temperature up in the winter and lower it during the summer, making for more comfortable surroundings all year round. Rugs and wall hangings can also help.

Low-emissivity, or Low-E glass, is often used for new windows because it makes the interior more comfortable. Emissivity refers to a material's ability to give off, or radiate, heat in the form of long-wave infrared energy. In the winter the Low-E coating on the glass reflects radiant heat back into the house rather than letting it escape outdoors. In the summer, the coating works in reverse, deflecting incoming heat away from the glass.

Personal Factors

The number of people who are going to occupy a building and their level of activity are important considerations in designing an HVAC system. A person who is working or playing hard generates about 600 watts of heat.

A decrease in air temperature can be offset by an increase in the radiant temperature or vice versa. People may not mind a lower air temperature in the winter if the velocity of the air is reduced or the relative humidity is increased.

HVAC Controls

HVAC system functions provide a controlled environment in which the following parameters are maintained within desired ranges: temperature, humidity, air distribution and indoor air quality. The control system must be able to directly control the first three parameters. Indoor air quality is influenced by the first three parameters but it may require separate controls.

Temperature Control

Temperature control is often done by a temperature controller called a thermostat which is set to the desired temperature value or setpoint. The temperature deviation, or offset, from the setpoint causes a control signal to be sent to the controlled device.

In a chilled water or heated water coil, the temperature controller may position a water valve to vary the flow rate of the heated or chilled medium through the coil. It can also control dampers at the coil to vary the proportion of air passing through the coil.

The valves used to control the water flow through the coil may be either two-way or three-way. The valves used to control steam flow through a coil are usually two-way.

Humidity Control

Humidity control in a conditioned space is done by controlling the amount of water vapor present in the air in the space. When the relative humidity at the desired temperature setpoint is too high, dehumidification is needed to reduce the amount of water vapor in the air. When the relative humidity at the desired temperature setpoint is too low, humidification is used to increase the amount of water vapor in the air.

The relative humidity depends on the dry bulb temperature. So dry bulb temperature and relative humidity are usually specified together. In a room at 70°F dry bulb and 50% RH, the actual moisture content (specific humidity) is 54.5 grains of water per pound of dry air. A room with the same specific humidity at 60°F will have about 70% RH and 80°F, this will be about 35% RH. Humidifiers use water spray, steam grid and steam pan techniques.

A humidity controller is known as a humidistat. It is usually adjacent to the thermostat so the humidity will be based on the ambient temperature. The humidity controller is set at the relative humidity setpoint. A change in relative humidity from the setpoint causes a control signal to be sent to the controlled component.

If a duct-mounted steam grid humidifier is used, the control signal will open the steam valve at the inlet to the humidifier unit. When the steam valve is open, steam flows through the humidifier in the supply airstream, which raises the relative humidity.

Another humidity controller may be located downstream from the humidifier to act as a high-limit safety control. If the relative humidity of the airstream reaches the saturation point, the high-limit control will reposition the steam valve and decrease the steam flow. This is done to prevent condensation and water flow downstream from the humidifier. An electrical heater steam humidifier will be controlled with electric contactors.

Dehumidification

Typical dehumidification methods include the use of desiccants, surface cooling and indirect cooling coils. Surface dehumidification is done on cooling coils simultaneous with sensible (dry bulb) cooling. Indirect cooling coils are used with sprayed coil dehumidification.

Dehumidification is usually done with sensible (dry bulb) cooling by surface dehumidification on the cooling coils. Indirect cooling using chilled water or another heat transfer medium can be used. Direct expansion refrigerant evaporator coils may also be used. Dehumidification in low dew point process systems is usually done in a separate dehumidification unit.

Air leaving the cooling coil during surface dehumidification is usually near a saturated condition. If cooling is controlled from the relative humidity in order to remove water vapor, the supply air will often be cooled more than is required for sensible or dry cooling. The space may require some reheating to prevent overcooling.

Intelligent HVAC

One of the most cost-effective improvements that can be made to an older HVAC system is to increase its intelligence. A setback thermostat, which is also known as a clock or electronic thermostat, will automatically raise or lower indoor temperatures to fit the building's schedule.

Indoor temperatures can be setback to 60°F while the building is unoccupied, then automatically brought back up before it is fully occupied. Another setback might start at or before noon when the outside temperature starts to warm.

Without making sacrifices in comfort, a 10 to 20% savings on the annual heating bill is possible. In the summertime the setback can help save on air conditioning. Some clock thermostats have 7-day programs and hold a different program for every day of the week. A slightly less flexible unit is the 5 + 1 + 1 thermostat, which has a single Monday-through-Friday program, with separate programs for Saturday and Sunday. There are also 5 + 2 models, which hold two programs, one for the five workdays and one for the weekend. The least flexible are the 1-program models.

These programmable units range from electronic units with a few buttons to electromechanical model that uses small levers or pins to mark the setbacks. A manual override lets you manually override the current temperature setpoint without reprogramming the thermostat. Some models automatically revert to the original program after the override period has ended.

Most electronic thermostats have a key or switch that allows you to change from a heating season program to cooling, or back again without reprogramming. Some models will make the changeover automatically.

Most electronic thermostats use batteries to hold the program in case of a power failure, but some use batteries as their only source of power. Thermostats without a low-battery indicator may fail without any warning.

Heating-only and cooling-only systems have two or three wires, and combined heating-cooling systems typically have four or five wires. Heat pumps use eight or nine wires.

Most units use a 24-volt power source, but there are also 120-volt or millivolt systems.

The thermostat should be about 5-feet off the floor on a wall that is not facing unconditioned air space or that has pipes or flues in it. The unit should not be near air outlets, radiators, drafty spots, in a closet or behind a door. The location should also be free from direct sunlight.

Another key improvement is to equip systems with more and better sensors so that the central controller has the feedback it needs to make smarter adjustments. These sensors monitor not only outdoor-air temperature, but relative humidity and wind conditions as well. It is the climate that defines heating and cooling needs so current weather information is needed. With the concept of zoning becoming more popular, control systems will have several subordinate thermostats and/or remote sensors throughout the building.

Dry Air Problems

Unless a forced-air system is in balance, the supply of warm air will exceed the cool air flowing out. Some areas will be pressurized while others are depressurized. The depressured areas will be invaded by cool, dry air from outside. If there are ductwork leaks, the depressurization effects and dryness become worse. If the ductwork is used for air conditioning, the depressurization can pull hot, humid air inside.

Even if the supply and return air in the system are well balanced and the ductwork is tight, the furnace may draw its combustion air from indoors and cause depressurization. In order to reduce the dryness and fuel consumption you can raise the relative humidity inside. Another solution is to add new supply or return air ducts for a better balance or regulate the system by opening and closing air outlets.

Tightening the exterior envelope will slow the passage of cold, dry air from outdoors to in and raise the relative humidity level inside. You feel warmer when the relative humidity goes up, even if the air temperature has not changed.

Providing the furnace with more outdoor combustion air will stop the furnace from depressurizing the rest of the building and eliminates the danger of backdraft. This also saves energy, since the furnace is no longer using previously heated air for combustion.

Forced-air heating has another comfort problem. The air may be forced out at 90°F or warmer over people's skin leaving them feeling cool and dry. Moving air causes convective heat losses [cooling] from the surface beneath it and also increases the evaporation off the skin.

This discomfort from moving air is more pronounced with heat pumps, which usually deliver air at 90 to 110°F. Most oil and gas furnaces have outlet air temperatures that are fifteen or twenty degrees higher. Warm air can hold more moisture than cold so the moving air also has a drying effect.

Zoning

When heat is delivered only to those areas where it is needed and when it is needed, waste is kept to a minimum. Early attempts to zone forced-air systems produced systems that were hard to balance and control. The advent of low cost microprocessor controls and motorized dampers to modulate airflow has allowed forced-air zoning to be practical.

Systems such as Carrier's Comfort Zone are designed for forced-air systems that have a single furnace or heat pump. It makes cost-effective zoning possible for areas as small as 1,400 square feet. The system creates up to four zones, each with its own thermostat. Sensors in the ductwork report on the system's status so that temperature, humidity, and airflow adjustments can be made. There is also an outdoor sensor so that the system can react to changing weather. According to Carrier, the system should pay for itself in saved energy in about 3-1/2 years.

The Harmony forced-air zoning system, from Lennox, can also control four zones and will provide 3- to 5-year payback. The Honeywell/Tril-A-Temp system can handle up to three zones.

Oil and Gas Furnaces

Furnaces are often categorized by their efficiency. This is a measure of how much usable energy they can deliver from each unit of fuel. The U.S. Department of Energy tests and rates oil and gas furnaces, assigning each model an Annual Fuel Utilization Efficiency (AFUE). Furnace performance can also be measured by combustion efficiency. AFUE ratings can range from 78% efficiency to the high nineties. Furnaces with efficiencies of about 90% are known as high-efficiency or condensing furnaces. Installation costs of a high-efficiency furnace are about 25 to 30% more than a conventional model. Since they burn less fuel, these high-efficiency units are a better investment over time, especially for larger areas.

High-efficiency furnaces use two separate heat exchangers. The hot combustion gases that would ordinarily be lost are forced through a loop of pipes (the second heat exchanger) where they give up more heat. The hot gas cools enough where water vapor condenses from it. So, these are called condensing furnaces.

The heat exchangers in condensing furnaces are built with special types of stainless steel and other alloys that can stand the hot, corrosive environment that is created. A condensing furnace can produce several gallons of water a day, which is drained out of the heat exchanger into the sewer line. The pH ranges from 2.4 to 4 with an acid level comparable to lemon juice. The drain line is sometimes equipped with a small lime-filled cartridge to neutralize the acid.

The high-efficiency models drop the gases coming out of the furnace to around 100°F. Instead of water vapor coming out on top of the chimney, it cools and condenses inside, depositing acid on the chimney liner. The flue gas temperature is so low that a special type of plastic pipe can be used for the flue.

High-efficiency furnaces and some of the lower-efficiency models draw their combustion air from the outside. There is no danger of backdraft and heated air is not used for combustion. Depressurization is less so there is less cold, dry air pushing its way inside. The furnace's heat exchanger can last longer too, since damaging indoor air pollutants like fumes from cleaners and aerosols are kept out of the combustion air. When these types of airborne pollutants get into a condensing furnace, they can form hydrochloric and hydrofluoric acid, which can eat holes in a heat exchanger.

Factors to consider in a forced-air system include:
- Variable-speed blower – allows the system to come up slowly and quietly, ramping the fan speed up or down as demand calls,
- Humidifier – combats the dryness problem,
- Electronic air cleaner – improves the environment as well as the life of the HVAC system.

Heat Pumps

Heat pumps use a reversible refrigeration cycle so they can produce either hot or cool air. Both heating and air conditioning are done with one compact unit. They are all-electric, so they do not need a flue or chimney.

Since heat pumps act as both heaters and air conditioners, the Department of Energy assigns them two separate ratings. There is the Seasonal Energy Efficiency Ratio (SEER) which is a measure of cooling. Typically, SEERs run from 10 to 17. The other rating is called the Heating Season Performance Factor (HSPF) which is a measure of heating efficiency. These range from 6 to 10. In cold-weather climates, as air temperatures fall toward freezing and below, air-to-air heat pumps have to use a back-up heater, which is usually an electric resistance coil built into the airhandler. There are also ground-coupled heat pumps which draw heat out of the ground. These are more effective in cold climates. In areas with a more severe winter than summer, a model with a high HSPF may be called for. If there is not as much of winter, but a lot of air conditioning days, a high SEER unit is best.

Electric Furnaces and Combo Units

In addition to gas- and oil-fired furnaces and electric heat pumps, there are other choices for powering a forced-air heating system. One of these is an electric furnace, which is made up of electric resistance coils packed into an airhandler. Due to the high cost of electricity in many areas, electric furnaces are being replaced with more economical units.

One alternative is a combo system or integrated system, as they are sometimes called. These are a combination of hydronic and forced air. Hydronic refers to the use of hot water or steam in delivering heat.

Combo units use a high-efficiency water heater or boiler, which serves as the heat source for both hot water and space heating. Hot water is pumped out of the water heater or boiler, through an insulated pipe, and into the airhandler. In the airhandler, the hot water winds goes through the heat exchanger coils. Air is forced over the hot coils and distributed through the building. After the water leaves the airhandler coils, it is circulated back into the water heater or boiler for another exchange loop. Some integrated systems use a boiler as the power source and are designed with separate boiler water and potable water lines.

Capacity Sizing

If a high efficiency system is to deliver its potential energy savings, it must be sized correctly. The most common installation mistake dates back to the dawn of the HVAC industry. This is putting in equipment that has too much capacity for the job. This practice of oversizing equipment may have gotten its start before the Civil War, when people believed in leaving a window open since stale air was suspected to cause many diseases. Early contractors made it a practice to oversize heating systems by about 40% to compensate for the open windows.

Many HVAC contractors specify oversize by a wide margin so they won't have any low capacity problems. Oversized equipment not only costs in equipment, but instead of running efficiently for long periods, it tends to cycle on and off using extra fuel. Oversized equipment also tends to be noisy. Sizing a heating system properly requires the following:

- The outside design temperature. This is the wintertime temperature extreme that the heating system will be designed to handle.
- The heated volume of the space.
- The surface areas and R-values of the floors, walls, ceiling, windows, and door.
- The average wintertime infiltration rate.

The infiltration rate is expressed in Air Changes per Hour (ACH) which is the number of times the total air volume turns over in an hour. For example, if the space volume is 10,000 cubic feet and the infiltration rate is 12,000 cubic feet per hour, the space would be rated at 1.2 ACH. This can be estimated by inspecting the building and categorizing it on a standardized table, but the infiltration rate can be more accurately estimated by testing the house with a blower door.

Hydronic Heating Systems

Hydronic or hot water heating systems tend to distribute heat more gently, with fewer drafts. Forced-air systems may leak and operate out of balance, which pressurizes or depressurizes different areas. This can draw cold, dry air inside affecting both comfort and economy. Hydronic systems can get out of balance but there is no infiltration problem due to depressurization.

Hot water systems allow extra precision and control. A hydronic system has to be free of leaks, but a forced-air system can lose a lot of heat between the furnace and the heated area. The precision with which hot water systems can be controlled makes them easier to zone. Thermostatically controlled radiator valves allow even old radiators to be zoned.

Hydronic systems also take less space since the heat-carrying capacity of water is 3,000 times greater than it is for air. Hydronic systems use small-diameter pipes to distribute heat, while forced-air systems must use large ducts. Most modern radiators are slim and flat, designed to save space. Hydronic heating systems that have underfloor distribution do not take any useable floor space.

Hot water heating systems except for steam are quieter than forced air since there is no blower turning on and off all the time and noise going through the ductwork.

Hot water heating systems typically cost more to install than forced-air systems. They also tend to be slower to respond to a change in the thermostat than forced air and cannot be easily adapted to handle air conditioning or ventilation.

If the pipes freeze or a leak occurs, there can be major damage. Leaks in forced-air systems usually cause only comfort problems, noise, and higher fuel usage.

Hydronic systems use a central boiler or water heater. The resulting hot water or steam is then distributed through a piping system to radiators which release the heat. When the water or steam has given up its heat, it returns to the boiler for reheating.

Early steam systems operated without pumps and controls, but required someone to keep them fueled and maintain water in the boiler. The steam generated in the boiler would rise through the pipes under its own pressure into the radiators.

There was a thermostatic air vent on each radiator that remained open, the temperature rise was enough to close the vent. As the steam inside the radiator cooled and condensed, the water would flow through the same pipe in a one-pipe system or drain down through a second pipe in a two-pipe system. The pipes were pitched to allow gravity to return the water to the boiler.

Although these steam systems are noted for their simplicity, they have several weaknesses. The steam boiler has to raise the temperature of the water to 212°F before the system can start to deliver heat. The large volumes of heat delivered can result in large swings in indoor temperature. Steam systems can also be noisy if they have not been properly installed and maintained.

Gravity hot water systems are similar to steam, both in the way they look and operate. Like steam, hot water rises from the boiler into the pipes and radiators. As the hot water releases its heat to the radiator, it cools and drops back into the boiler. Steam boilers need to be partly full of air to work, but hot water systems are completely filled with water, except for the air that is left in an expansion tank. Since hot water systems do not have to boil to work, they begin to deliver heat to the radiators at lower temperatures. Hot water systems can be mechanically pumped, while steam relies on natural forces.

Many gravity systems were changed over to fuel oil or gas. The large pipes and radiators that were designed to handle slow-burning coal produced a more powerful heating system that was more difficult to control.

Along with the use of oil and gas boilers that fire automatically, the newer hydronic heating systems employ pumps to push hot water out of the boiler, through the radiators, and back into the boiler again. These are known as forced hot water systems.

The use of circulating pumps allows the hot water system to respond more quickly to the thermostat. It also permits the use of smaller pipes and radiators. The gravity system needed larger pipes and radiators so that water or steam moving through the system would not be slowed by friction. The boiler can also be located in other areas than the basement.

Hot water heating systems use a control device called an aquastat which control the temperature of the water. Instead of a water temperature set at 160 to 180°F, if the aquastat is set at 120°F to 140°F, you can save 5 to 10% on fuel. An outdoor temperature sensor on the heating system can result in even more savings and setbacks can be made automatically.

Hydronic System Controls

Hydronic systems with centrifugal pumps are controlled in a similar manner to air distribution systems with centrifugal fans. When the demand for heating decreases, modulating valves at heating elements are controlled to reduce the fluid flow and the piping pressure increases as the pump moves on its characteristic curve. To prevent system damage from pressure buildup and achieve energy savings, the pump speed is varied to maintain the piping system pressure within acceptable limits.

Hydronic Piping Systems

Hydronic piping systems serving a number of modulating control valves, such as found in a fan-coil unit system, may use the following alternatives:
- Use 3-way valves and operate as a constant flow pumping system,
- Use 2-way valves and operate as either constant flow or variable flow system,
- Use a mixture of 2-way and 3-way valves and operate as a limited variable flow system.

Flow Rates

A constant flow system is needed in a chilled water piping system because, the chiller water flow should be relatively constant for the control strategy to function. All chillers have a minimum water flow to prevent freezing of low water flows through the evaporator tubes.

Some boilers have a similar minimum flow requirement to prevent hot spots on tubes and to minimize the thermal shock from cold return water. The minimum flow must be enough to prevent an excessive increase of fluid temperature in the system due to pump heat produced by the pump motor operating at near shutoff head.

In systems with a mixture of 2-way and 3-way valves which operate as a limited variable flow system, one or more 3-way valves are selected to give a flow rate equal to the minimum flow rate. These are used at the ends of piping loops with 2-way valves used for closer-in loads. The 3-way valves provide a minimum system flow while the 2-way valves modulate and increase the system pressure.

Some hydronic systems are built so tightly that they continue to cycle year after year without losing any water from the system or collecting much sludge in the boiler. Others need to have water added frequently and accumulate large amounts of sludge.

Even in the summertime, it is recommended that the boiler be fired for awhile. Heating up the system will drive air (oxygen) out of the fresh water and help prevent rust.

Automatic Shutoff and Refill

As water and sludge are drained out, the boiler should automatically shut itself off. The automatic shutoff is a built-in safety feature designed to protect the boiler against low-water conditions.

Many systems have an automatic refill control, it should automatically refill the boiler to its proper level. Since water expands when it is heated, hot water heating systems have a special tank so that the water can expand in a safe and controlled fashion.

Expansion Tanks

Early expansion tanks were located somewhere above the highest radiator, usually in the attic. The top of the tank was open, allowing air to move in and out of the tank as the water level and temperature in the system changed. The next tanks were steel cylinders that were typically hung above the boiler. These tanks were designed to operate under pressure, half full of water, half of air. One problem with this closed tank arrangement is that the pressure forces air into the water. When that water gets into the radiators, the pressure and temperature drop and air gets into the radiators. As the radiator fills with air, it loses efficiency and can be noisy.

Modern expansion tanks use a neoprene bladder that expands or compresses with the changing volume of the water. The water and air do not mix. These newer expansion tanks do not need to be drained, while most of the older models units have a drain valve and should be flushed out once a year.

When air gets trapped inside a radiator, it prevents it from filling completely with hot water and the heat output is low. This must be resolved by bleeding the air out of the radiator.

There are also fixtures that can be installed in the piping near the boiler. The water passing through it is forced through a tightly wound spiral of wires that forces air bubbles out of solution and vents them to the room.

There is no need to bleed the radiators in a steam system since they should have air in them, but steam radiators do have air vents that control the flow of steam. These vents clog up making them hiss and spit steam.

Water Hammer

Pipe insulation will save fuel and also reduce or eliminate water hammer. This is a loud, clanging noise that cold pipes can make when hot water or steam surges suddenly through them. Insulating the water pipes returning to the boiler will prevent them from dripping from condensation.

In a steam heating system, compressed fiberglass insulation should be used on the pipes rather than foam. Steam pipes can become hot enough to melt foam insulation. Many older steam systems were insulated with asbestos, which is a respiratory hazard if it breaks loose and is dispersed in the air. If pieces are broken or missing, contact a company that specializes in handling asbestos.

Radiators

Radiators come in many sizes and shapes. Older units were usually cast-iron. The newer European-style radiators are made out of aluminum, copper, and high-temperature plastics.

Most radiators use a combination of convection and radiation to heat a room. Convection is the transfer of heat from the hot metal surface to moving air. Radiation is the transfer of heat by electromagnetic waves, which move from a warm object to a cooler one without noticeably warming the air in between.

The radiator's design determines how much heat is released through convection and how much is given off through radiation. Baseboard radiators that have a copper pipe ringed with fines (finned-tube heating) are designed to heat the room mostly by convention. The fins increase the surface area of the radiator and promote air movement.

Low-temperature panel radiators use a plastic or metal water tube encased with steel or aluminum. They deliver most of their heat by radiation. The best color for heating efficiency is dull black.

Radiant Floors

In-floor heating use a serpentine loop of hot water lines that are either embedded in the cement slab when it is poured or looped through subfloor joists. This turns the floor into a radiator.

Early in-floor systems used a continuous loop of steel or copper pipe, which was embedded in the concrete. The controls were inadequate and did not maintain a continuous and even flow of water through the slab. The controls would cycle the boiler on and off, sending a surge of hot water through the slab then letting it cool. The pipes in and out of the slab were subject to stress and often failed. The lime in the cement also tended to corrode the copper and steel pipe. Many of those early in-floor hydronic systems suffered from premature leaks and had to be sealed off and replaced.

In the late 1960s new plastics developed for telecommunications lines found an application in hydronic heating. Cross-linked polyethylene retains its properties under wide and repeated swings in temperature. It also has a diffusion barrier that prevents oxygen from passing through it. This protects the pump, valves, and fittings in the heating system from corrosion. Although plastic tubing does not transfer heat into the floor as well as copper or steel pipe, some tests indicate that cross-linked polyethylene can last 200 years inside a slab. Many installations have been in place without a failure since 1969. The control system should supply the slab or subfloor with a steady supply of hot water. Typical circulation temperatures are 80 to 130°F, which is cooler than other types of hydronic systems. The control system should have an outdoor temperature sensor to help the system anticipate the heating needs.

On new foundation work the polyethylene tubing for the in-floor heating system is laid out and the concrete slab is poured over the top of it. The poly-ethylene tubing can also be laid out on the top of the wooden subfloor and a slurry of lightweight concrete poured over it.

Another approach is to run 8-inch wide sleepers on an existing floor with the loops of polyethylene tubing between the sleepers. When it is impractical to pour a new concrete floor or use sleepers to raise the floor, the tubing can be installed on the underside of the floor. Holes are drilled through the joists so that the polyethylene tubing can be threaded through them. In Germany in-floor heating accounts for 60 to 70% of all new systems installed.

Comfort Solutions

One of the common problems with older hydronic systems is that they often operate out of balance, producing hot and cold spots in different areas. Sometimes this can be resolved by adjusting the radiator valves.

The valves cannot be adjusted in a steam system, since the radiator valves that control incoming steam must be wide open to work properly. The size of a radiator's air vent can be changed. A vent with a larger orifice allows steam to enter the radiator more quickly, providing faster heat.

A better solution is to equip each radiator with its own thermostatically controlled valve. This turns each area around a radiator into a separate temper-ature zone that can be individually controlled. Marked improvements in both comfort and economy are possible. Thermostatic radiator valves can be used on steam or hot water systems, as long as the radiators are not connected in series.

Weather-responsive Controls

Another common problem with hot water systems is that they are slow to respond to changing demands. The solution is to use a weather-responsive con-trol. An outdoor sensor, which works with the aquastat on the boiler, allows the system to anticipate rising or falling demands for heat. Weather-responsive con-trols, which are also known as modulating aquastats or reset controls, can save up to 25% on fuel.

Boiler Ratings

Boilers are tested and rated by the federal government. Each model is assigned an Annual Fuel Utilization Efficiency rating (AFUE) that estimates the boiler's annual performance. New boilers have AFUE ratings that run from 80 to 90%. The condensing models are designed to extract more heat from the fuel, cooling the combustion gases to the point where water condenses in a heat exchanger and flue.

Depending on the temperature of the flue gases coming out of the boiler, it may be possible to use plastic flue pipe, such as Plexventor Ultravent, which can be piped out through a side wall. The boiler should be sized to meet the load and not exceed it. A properly sized boiler will run almost continuously in very cold weather.

Boilers that were developed back in the 1970s use a small amount of water as a response to the energy crisis. These units cycle on and off so much that the equipment becomes easily fatigued. The recommended minimum cycling time for a hot water boiler should be five to six minutes.

Direct Heating Systems

In direct heating, no ductwork or pipes are needed to distribute the heat. Direct heating includes baseboard electric resistance heating and radiant electric.

Baseboard electric radiators use a heating element that is wrapped in a tubular casing with metal fins along its length. When electricity flows into the resistance heating element, it produces heat. The fins increase the heated surface and promote the movement of air up through the radiator.

Electric baseboard radiators come in 1- to 12-foot lengths, with ratings of 100 to 400 watts per foot. They are placed around the outside wall of the room and positioned under windows to counteract the infiltration, conductive heat losses, and drafts associated with the window glass and frame.

Electric baseboard radiators are easy to zone. Some models have a thermostatic control built into the radiator while others are designed to operate with wall-mounted thermostats. With baseboard electric there is no need for a chimney or flue, there are no problems with backdrafting combustion gases or depressurization.

Some electrical companies offer a special meter along with special rates for electric resistance heating. In a climate where the winter is mild, or where there are special electric rates, baseboard electric can be practical as a heating system.

Radiant Electric

Instead of using baseboard radiators to convectively heat the air, radiant electric systems are designed to heat objects. Radiant electric heat can take place with flexible element heaters above the ceiling or underneath the floor. These are prefabricated sheets with a metallic or carbon-graphite conductor sandwiched between two layers of insulating plastic. The conductor may also be printed directly on the plastic. As the element becomes warm, the heat spreads into the ceiling or into the subfloor and flooring. These areas then radiate heat to the objects inside the room.

It is necessary to insulate the cavity between the joists so that as much heat as possible is directed into the room. Then the flexible element heater is stapled in place and the ceiling or subfloor goes in. Some types of flexible element heaters are factory pre-cut while others are delivered in rolls and cut at the site. Widths range up to 48-inches. Flexible radiant heating panels are sold under the following brands: Flex-Heat, Flex-Watt and Aztec Flexel.

Another technique in radiant electric heating for ceilings involves a special type of gypsum board with nichrome heater wires embedded in its core. This type of product includes the Panelectric and Suncomfort boards. These gypsum board heaters are heavier and less flexible to install than plastic sheets. They are relatively inexpensive, however, and also provide the finished ceiling.

A grid of special cable can also be embedded directly in the concrete or masonry of the floor. Raychem's RaySol in-floor heating systems use a special polymer in the electric cables that allows the system to regulate itself.

Radiant electric heat is also available in modular panels, which can be mounted against an existing wall or ceiling. The panels come in various sizes, wattages and colors. These modular panels from Enerjoy and Aztec tend to be more costly than other radiant electric systems, but since they do not have to heat up the ceiling before they deliver heat, they respond faster.

Radiant electric heat is easy to zone, does not take up any floor space for radiators, and there is no maintenance. In some applications, such as rooms that are only intermittently heated, rooms with very high heat loss, or for heating a small part of a large room, radiant electric systems can cost less to operate than electric baseboard radiators and other warm-air systems. The system should have a limit switch that automatically breaks the circuit if the unit starts to overheat.

Environmental Risks

Radiant electric heating systems generate an electric field which may have possible health risks. The effects of the electric field depend on the amount of current (amps) running through the system, how close you are to the field and how often you are exposed to it, and the geometry of the wiring. Systems with large open loops set up a stronger magnetic field than systems that route the current back along the same line. Other variables, such as how often the system cycles on and off, can also be important.

AC fields create weak electric currents in the bodies of people and animals. This means there is a possibility of biological effects. These currents from electric and magnetic fields are distributed differently within the body. The amount of current is very small and too weak to penetrate cell membranes but it is present between the cells.

Some studies have shown an increased cancer risk for workers around electrical equipment. About 50 studies have reported statistically significant increased risks for several types of cancer in occupational groups that have elevated exposure to electrical and magnetic fields.

Although the use of electricity has increased greatly exposures to these fields have not increased in the same way. There have been changes in the way that buildings are wired and electrical equipment have become smaller and more efficient which has resulted in field levels. Rates for various types of cancer have shown both increases and decreases over the years.

Several studies have concentrated on the effects on pregnancy outcomes and general health. The sources for possible association with miscarriage risk have included electric blankets, heated water beds and electric cable ceiling heat.

Some studies have correlated exposure with higher than expected miscarriage rates while others have found no correlation. Epidemiologic studies have revealed no evidence of an association between exposure and birth defects in humans.

A 1979 study reported a possible link between electromagnetic fields and childhood leukemia. This report was followed by additional epidemiological studies, some of these reported similar links between low-frequency electromagnetic fields and cancer, but others found no association.

A report by the Oak Ridge Associated Universities concluded there was no convincing evidence that exposure to these fields is a demonstrable health hazard. A similar conclusion was reached in a review conducted by the United Kingdom's National Radiological Protection Board.

A case study of cancers in children living near high tension power lines in Sweden found a link for childhood leukemia. No association was found for other childhood cancers. In this study, children living within 50 meters of high tension power lines were found to have almost 3 times the risk for childhood leukemia. According to the study, of the 80 or so childhood leukemia cases diagnosed in Sweden each year, power lines may be responsible for one. Other studies have been planned to check these results. The U.S. Environmental Protection Agency has an Electromagnetic Field Hotline at 800-643-4794 which is a good source for up-to-date information.

Air Conditioning

Central air conditioners use a compressor, two coils, and a chemical refrigerant to extract heat from the indoor air and transfer it outside. As the liquid refrigerant moves through the indoor coil (the evaporator), it absorbs heat from the air and changes it into a gas. The gas then moves through the compressor, which compresses the gas and raises its temperature. The gas is sent to the outdoor coil, the condenser, where it releases its heat to the outside air and condenses back into a liquid.

Air conditioning cools in two ways:

• It removes heat energy from the air and lowers its temperature,
• It removes water vapor from the air.

In addition to cooling and dehumidification, a central air conditioning system will also control air circulation and ventilation, and have some means to clean the air.

Heat pumps are a special type of air conditioner that can be reversed in winter to provide heat. During the summer, the heat pump's compressor and refrigerant loop function like a standard air conditioner, drawing heat out of the indoor air and ejecting it though the outdoor coil. In the winter, the process is reversed and the heat pump absorbs heat from outdoors and sends it inside. At 0°F, the air still contains more than 80% of the heat it held at 100°F.

The heat pump's heating efficiency is affected by low temperatures. As the heat pump loses some of its efficiency, an electric resistance heater built into the airhandler is switched on. Heat pumps are generally used where the air conditioning load is much greater than the heating load.

Geothermal Heat Pumps

A newer type of heat pump can provide excellent heating and air conditioning efficiency regardless of the outdoor air temperature. These are geothermal heat pumps (GHP) which may be open or closed loop.

Geothermal heat pump operation is based on the fact that the earth remains at a relatively constant temperature throughout the year, warmer than the air above it during the winter and cooler in the summer, like a cave. The GHP takes advantage of this by exchanging heat with the earth through a ground heat exchanger, using an open- or closed-loop system.

The open-loop type is also called a water source heat pump. It must have a water source like a well, pond or river nearby to operate. An open-loop system has a water supply pipe and a discharge pipe. The heat pump is equipped with a heat exchanger that draws heat out of the water in the winter and dumps heat into the water during the summer.

This type of system is more efficient than an air-to-air heat pump because the geothermal properties of the earth keep the ground water between 40 and 60°F in any season. The efficiency can be less when lake or river water is used, since the water temperature from those sources may be higher and not as constant as groundwater.

Closed-loop geothermal systems are called ground-source or ground-coupled heat pumps. This type of geothermal system has a continuous loop of polyethylene or polybutylene pipe that is buried in the ground. The depth and arrangement of the piping laid depend on the size of the heating and cooling load and the geological features at the site.

A heat transfer fluid (water mixed with antifreeze) is circulated through the ground loop into the heat pump's heat exchanger and then back down again. In summer, heat is drawn out of the house and transferred into the ground; during the winter, heat is drawn from the ground and transferred to the airhandler. No matter how hot or cold the outside air temperature gets, the earth around the ground line stays at about 50°F.

While a geothermal heat pump costs more initially than other heating and cooling systems, it compensates for this with lower operating costs. In some areas, utilities will help pay for a geothermal installation. Geothermal heat pumps include Water-Furnace, Florida Heat Pump, and ClimateMaster.

A Geothermal Heat Pump (GHP) system is designed to offer year-around comfort and economical operation. It can allow savings of as much as 20 to 50%. Other benefits can include reduced maintenance costs, improved heating and cooling reliability, and conservation of natural resources. According to the International Ground Source Heat Pump Association, GHP investments can be recouped in as little as 3 years. In a study by the Red River Valley Rural Electric Association in Oklahoma, 99% of the GHP owners rated the heat pump's ability to provide summer comfort as an 8 or higher on a 1-to-10 scale.

There are several different ways to bury the plastic pipe loops. The method chosen will depend on the available land area and the soil and rock type at the installation site. The most common approach uses a closed-loop system of high density polyethylene pipe. Pipes are buried horizontally at 4 to 6-feet deep or vertically 150 to 250-feet deep, and filled with a water solution. In the winter, the fluid within the pipes extracts heat from the earth, carrying it into the building. In the summer, the system reverses itself. Heat is pulled from the building, carried through the pipes, and deposited in the cool earth. A fan located inside the building distributes the warmed or cooled air throughout the interior.

Horizontal installations, when a trencher or backhoe can be used, are less expensive, but take up more land area. Vertical installations, where well drilling equipment is used, are generally more expensive but are ideal where land is scarce. Retrofit installations, where there is an existing building, can use horizontal boring. This method disturbs the least amount of land. An open loop system can be installed where an adequate supply of suitable water is available and open discharge is feasible.

The GHP uses electric power only to move heat. During summer cooling, the heat from inside the building is returned to the earth with the same high efficiency. Another advantage comes from configuring the system so the waste heat from air conditioning is used to provide free domestic hot water in the summer and at a substantial savings in the winter. Special financing may also be available through the local utility or through the producer of the system.

Another technology is the natural gas-fired heat pump provided by York International, which worked with the Gas Research Institute to develop the technology. A gas-fired heat pump can deliver air outlet temperatures that are 10 to 15°F warmer than electric air-to-air heat pumps and will operate more efficiently at lower outdoor temperatures.

Maintenance

Common sources of efficiency problems in heat pumps include:
1. Leaking ducts,
2. Dirty coils and air filters,
3. Too much or not enough refrigerant in the system,
4. Thermostat problems that cause the heat pump's electric backup heater to run too often or not enough.

One study in Oklahoma, conducted by the Alliance to Save Energy, U.S. Department of Energy, Oak Ridge National Laboratory, and Public Service Company of Oklahoma, found that the replacement of older inefficient window air conditioners with higher efficiency models can reduce air conditioning bills by one-third or more.

If air conditioning is added to a forced-air system, it is important that the ductwork be tightly sealed and insulated. This will prevent condensation from dripping off the ducts.

Ductless Air Conditioners

A split-system or ductless air conditioner is made up of a compact outdoor unit, with one or more compressors, which serve a matching number of indoor airhandlers. A special rigid conduit of 2- to 3-inches in diameter is run from the outdoor unit, through the wall, to an airhandler mounted inside. The insulated conduit contains a power cable, liquid and suction tubes for the refrigerant, and a condensate drain.

The indoor airhandlers can be wall, ceiling or floor mounted. Each airhandler has a small fan and is individually controlled for zoning.

A switch is used to select cooling or heating functions and a temperature controller cycles the cooling and heating functions. Heating may be from a reverse cycle heat pump or from an electric resistance heater.

Through the wall packaged terminal air conditioners (PTACs) are set in wall sleeves with plug-in electric connections for cooling and heating. There may also be piped hydronic connections for hydronic systems. The refrigerating section operates with integral safety controls. With hydronic heat the controller positions a solenoid valve at the hydronic coil.

The more powerful split-system air conditioners can run several separate airhandlers or several rooms. The efficiency of split systems in SEERs (Seasonal Energy Efficiency Rating) ranges from 10 to 12.

Plastic flex duct of 2-inches in diameter is also used. This plastic duct is corrugated which impedes the airflow and this type of system requires a higher velocity fan than other air conditioning systems.

High-velocity air conditioning systems can also be adapted to provide heating. This can be done in the following ways:

• Heat pump,
• Hot-water coil from a boiler or water heater,
• Electric heating element.

Hydrotherm and Unico are among the companies that have pioneered high-velocity air conditioning. Unico was founded in 1985 to manufacture a new and improved high-velocity air conditioning system without ductwork.

The high velocity system traces its roots back to 1946, when an engineer named Calvin MacCracken who worked on the development of the jet engine, installed a unique heating system in a house in New Jersey. The system was named Jet-Heet and took some of its basic design principles from the jet engine. Fourteen years later, there were over 2,000 similar installations.

The Jet-Heet line then added a line of air conditioning systems known as Jet-Cool. In 1964, Space Conditioning purchased Jet-Heet and the product line evolved into a product known as Space-Pak. In 1984, the Space-Pak line was sold to Hydrotherm.

The Unico and similar systems work on the principle of aspiration. A stream of air flows into the room through a 2-inch outlet, located in the ceiling wall, or floor, creating a suction around itself. From the air outlet in the room, 2-inch flexible sound absorbing tubing is threaded around, over and under the existing inner wall construction. The only other visible part of the system is the return air grille, which is usually less than 3-feet square.

HVAC Control Functions

The basic control functions of the HVAC system include starting airhandling fan motors, emergency system shutdown, positioning outside air dampers, positioning mixed air section dampers for economizer cycle cooling, providing seasonal changeover or enthalpy-based control, providing water valve control for chilled water cooling coils, controlling refrigerant flow in cooling coils, controlling airflow to air terminal units, heated water or steam valve control and energization of electric heating coils.

Temperature Control

Temperature control in an air conditioning system that uses air as a delivery medium may employ one of the following techniques:

- Constant volume, variable temperature – Control the temperature of air supplied to the space while keeping the airflow constant.
- Variable volume, constant temperature – Control the airflow rate while keeping the temperature constant for air supplied to the space.
- Variable volume, variable temperature – Control both the airflow rate and the temperature for air supplied to the space.
- Variable volume reheat control – Vary both the supply air temperature and flow rate where the airflow rate is varied down to a minimum value then the energy input to reheat coil is controlled to vary the supply air temperature.

Air Volume Control

When the variation of supply air volume is used to control the space temperature, the temperature controller may:

- Cycle the fan motor in an on-off sequence,
- Modulate the fan motor speed and
- Damper the airflow using volume control dampers.

In a fan-coil unit, the space temperature may be regulated by regulating the air flow rather than the water flow through the coil. The temperature controller can vary the fan speed using a speed controller or cycle the fan on and off.

In a Variable Volume (VAV) system, the supply air volume changes the temperature controllers on individual terminal units position modulating dampers. The central station airhandling fan operates continuously, although the fan performance may be varied to maintain a static duct pressure within acceptable limits. Fan performance control methods include using the fan's characteristics, inlet or discharge damper control, inlet guide vane control or fan speed control by mechanical or electronic systems.

Variable volume air distribution systems typically use pressure control for the desired setpoint. A pressure deviation from the setpoint value causes a control signal to be sent to the controlled device.

If the load decreases, dampers are positioned to reduce the supply air volume. The restriction to airflow from the closing of dampers causes an increase in duct pressure and causes the fan to operate on its characteristic curve to reduce the airflow volume. Pressure control is used in order to prevent a large increase in static pressure. The duct static pressure change is interpreted by the pressure controller which positions the controlled devices to reduce the fan output and bring the system pressure back toward the setpoint value. The controlled device to modify the fan performance may be an inlet or discharge damper actuator, an inlet guide vane actuator, a mechanical speed control device, or a variable frequency drive.

Air Quality

Control of the air quality may be done by the following methods depending on the degree of contamination:

- Dilution with outside ventilating air,
- Filtration of particulate matter with air filters,
- Filtration of gaseous contaminants with odor adsorbent or odor oxidant filters and
- Hooding of contaminating processes.

HVAC System Classifications

HVAC systems can be classified based on the medium which is used to transfer heat within the system. All-air systems perform all the conditioning processes with air. These processes include cooling and dehumidification, heating and humidification, along with air cleaning and air distribution.

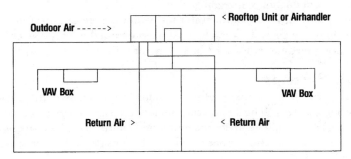

Figure 1-1 Typical Central System

Conditioning of the air is usually done at a central station. (Figure 1-1) An all-air system supplies only conditioned air to the space. No other cooling or heating medium crosses the boundary into the conditioned space.

In an air-water system, both air and water are used for cooling and heating. The conditioning of air and water is performed in a remote central plant and then distributed to terminal units in the conditioned space.

The chilled and heated water is delivered to the building from the central plant using a 2-pipe changeover or 3- or 4-pipe simultaneous piping system.

Air-water systems include fan-coil units with central ventilating air, terminal reheat units, induction reheat terminals, underwindow induction terminals and fan-powered induction terminals.

All-Water systems use heated or chilled water which is circulated through a terminal unit in the conditioned space. The terminal unit provides cooling and dehumidification or heating according to the zone requirements.

Conditioned air is not brought in from a central airhandling system. Outside air for ventilation comes from normal infiltration through window and door cracks or through wall intakes located behind each unit.

The terminal units can be fan-coil units or unit ventilators. The heating-only systems use heated water terminals, such as reheat coils and convectors. These are usually referred to as hydronic systems.

Packaged systems are similar to all-air systems since they perform the con-

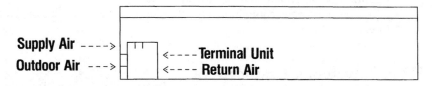

Figure 1-2 Typical terminal system

ditioning processes of cooling and dehumidification, heating, and ventilating with air; but, the apparatus is located in the conditioned space. Air distribution takes place from ducted or integral grilles. (Figure 1-2)

The conditioning of heating water for hydronic coils and heat pump loops is usually done at a central station located remote from the space. No other cooling or heating medium crosses the boundary into the conditioned space.

All-Air Systems

One of the most common all-air systems is the single-path, single-zone, draw-through system. The basic controller is a space temperature controller (thermostat) or separate temperature sensor and controller. When the supply fan is started, the control system is energized and the outside air damper opens to the minimum position.

The changeover of temperature controls between cooling and heating modes may be done either automatically or manually. The controls may allow for sequenced operation.

Temperature controls may be either 2-position or modulating. Two-position controls can be used to cycle a refrigeration compressor or position a refrigerant solenoid valve. Modulating controls can modulate a chilled water valve, heating coil valve or dampers.

In a 2-pipe changeover system where both chilled water and heated water are not available all year long, the heating system may be integrated with an economizer cycle to provide free cooling when available.

Safety Control

The fire and smoke safety control devices in all-air systems include code mandated devices such as smoke detectors, smoke dampers, manual fan shutdown switches and fireman control panels. The changes in some codes have reduced the requirements for fire safety thermostats or firestats, but many buildings still have them. The firestats or smoke detectors are installed in the supply air ductwork leaving airhandling units larger than 2,000 cfm. Systems with over 15,000 cfm capacity may have firestats or smoke detectors in both supply air and return air ductwork. These deenergize the supply fan and other interlocked fans when the air temperature reaches 125°F for return air and 165°F for supply air with electric or heated water heat sources. Systems with steam coils controlled by normally open valves are usually set at 300°F.

Duct smoke detectors may not be effective due to the dilution effects of the air being handled. Area smoke detectors installed in the space provide a more reliable means of smoke safety shutdown. When multi-floor return inlets are used, a separate smoke detector is required at each inlet. Smoke dampers may be required in some code-mandated locations.

Manual fan shutdown switches are usually furnished as breakglass stations similar to fire alarm boxes. These are required to be installed in exit pathways so that fans can be shutdown when the building is evacuated.

Fire control panels allow fire service personnel to restart fan motors stopped by firestats or smoke detectors and to position the dampers to evacuate smoke from the building.

Single-path, Multi-terminal Systems

Single-path, multi-terminal systems include variable air volume (VAV) single duct, ceiling induction reheat, constant volume reheat (CVR) and fan-powered terminal or powered induction units (PIU).

The central airhandling unit for a single-path, multi-terminal system is similar to the single-zone, single-duct system. The differences are that the supply air temperature and airflow volume are controlled. Space temperature is controlled by individual terminal units.

Variable Air Volume (VAV), Single Duct Systems

In single duct VAV systems the supply air temperature is held constant while the supply air volume changes to satisfy the cooling load. Terminal unit heating coils may be either hydronic or electric resistance.

As the volume of the supply air to the zones through the terminal units changes, the air volume delivered by the fan must also be adjusted. One type of control method uses motor-actuated variable inlet vanes on the fan positioned by a pressure controller that senses the supply duct pressure. The static pressure controller compares the static pressure in the duct with the pressure setpoint, determines the offset, and positions the variable inlet vanes to bring the duct pressure to its setpoint.

Another control method uses fan speed regulation. The air volume delivered by a centrifugal blower is directly proportional to its speed in rpm, while the pressure developed by the fan varies with the square of the fan speed, and the power required varies with the cube of the fan speed. The fan speed can be regulated by mechanical speed changes, frequency control and voltage control.

Induction Reheat (IR) System

These systems are usually mounted in ceiling plenums or under windows. The supply air temperature and pressure are held constant and the supply air volume to each terminal is changed to meet the load.

The supply air flows through an orifice that creates a low pressure inside the terminal unit. This induces a flow of room air from the return air plenum or from the space. In some systems, no heat source is provided in the central airhandling unit, but heating coils are provided in the terminal units. Damper control is used until the terminal has reduced primary air to the minimum flow rate. Then, the heating coil will be controlled to maintain the space temperature. The heating coil may be electric resistance or hydronic.

Constant Volume Reheat (CVR) Systems

In these systems air is supplied to the terminal units at constant volume and constant temperature. A reheat coil in each unit is controlled from a temperature controller. The air temperature supplied from the unit needs to be cold enough to satisfy the zone with the highest cooling load.

Powered Induction Units (PIU) Systems

In a PIU system, the terminal units are fan-powered mixing units with a supply air fan, a variable volume valve (VVV) and a heating coil. PIUs may be used for parallel flow with the PIU fan in parallel with the VVV or series flow with the PIU fan in series with the VVV. In both cases, the heating coil which can be either electric or hydronic, is the final element in the supply air stream. The parallel flow PIU acts as a variable volume/constant temperature unit at high cooling loads and a constant volume/variable temperature unit at low cooling loads and heating.

The series flow PIU acts as a constant volume/variable temperature unit. The PIU fan provides an almost constant volume supply of air.

Parallel-Path Multizone Blow-through Systems

Zone temperature controllers are used to position zone mixing dampers from cold and hot-decks. The total supply air volume remains almost constant. A cold-deck temperature controller may be used to position an automatic control valve on the cooling coil and the mixed air controls in an economizer cycle. Each zone temperature controller resets the cold-deck controller to supply the highest cold-deck temperature that will satisfy the zone with the greatest cooling load.

In systems with hydronic hot-deck coils, a hot-deck temperature controller positions a control on the heating coil. The setpoint is usually set in reverse sequence using an outside temperature sensor to increase the hot-deck air temperature as the outside air temperature decreases.

Tuning for hot-deck control is required to minimize energy wastage from the mixing of cold-deck and hot-deck airflows. Systems with electric hot-deck coils are subject to nuisance tripouts from high temperature cutouts when the airflow is reduced for light heating loads. A better electric heating solution is to install individual electric resistance duct heaters in the zone ductwork downstream from the mixing dampers. The zone temperature controls are used to energize the individual zone electric resistance heaters.

A firestat and smoke detector may be located in the return duct. An additional temperature controller may provide low temperature safety control to prevent coil freeze-up. This is done by deenergizing the supply air fan and closing the outside air dampers.

Dual-duct, Constant Volume Systems

In a dual-duct, constant volume (DD/CV) systems, the central station is similar to a multizone blow-through unit except mixing dampers are not provided. Cold-deck and hot-deck ducts extend from the unit to the conditioned areas.

The terminal units at each area have duct connections from the cold-deck and hot-deck ducts. Mixing of cold-deck air and hot-deck air takes place inside the terminal unit. The space temperature sensor or controller in each zone positions the cold duct damper and a constant volume regulator in the box controls the total air delivered. The control of total air volume supplied indirectly limits the amount of hot-deck air. Selected zones are used to set the cold-deck and hot-deck supply temperatures to minimize energy wastage from the mixing of cold and hot air.

Dual-duct, Variable Volume Systems

In these systems, a variable volume regulator is used on the hot-duct inlet instead of a constant volume regulator, or the hot-deck inlet is closed to allow the pressure dependent cold air damper to be the variable volume device. The latter technique may have some instability during start-up and during light loads when some cold-deck valves open and upset the pressure balance in the system; otherwise, it functions the same as the dual-duct, constant volume system.

Air-Water Fan-coil Units

These use central ventilating with 2-, 3- or 4-pipe systems. The temperature controllers are either wall-mounted or unit-mounted. A controller modulated water valve is used with constant fan operation although the occupant may select the fan speed setting. In an alternate control scheme, the controller cycles the fan or modulates the fan speed with a constant water flow through the coil.

In a 3-pipe system the airflow is manually selected as high, medium, or low. The temperature controllers are either wall-mounted or unit-mounted. They control the water flow through a 3-pipe valve at the combination cooling-heating coil. In the dead band between cooling and heating temperature setpoints, there is no water flow through the coil. There is no mixing of chilled and hot water. Water from the coils flows into a common return pipe. A 4-pipe system is more flexible and allows a low heating water temperature so it can be used with solar heat or reclaimed heat systems. The heating coil can be one loop deep. The separate cooling and heating coils can be built with separate tube bundles contained in the same fin bank. There is no mixing of chilled and hot water.

Fan-coil units with an outside air intake are similar to fan-coil units with central ventilating air except the ventilating air is provided through an outside wall. During subfreezing weather, a constant water flow through coils should be maintained to minimize the chance of coil freezeup.

Unit ventilators are fan-coil units with integral air-side economizer cycles. The components used in unit ventilators include a cooling/heating changeover switch or relay, room temperature controller, low-limit temperature controller, two-way valve, and motorized outside air/return air damper. Unit ventilators are often served by 2-pipe changeover systems.

Packaged terminal systems include unit ventilators with direct expansion cooling and electric or hydronic heat. Unit ventilators can be set in wall sleeves similar to PTACs or as a split system with an indoor unit set on a through-the-wall damper box and a remote condensing unit set at grade. Both system types use hard-wired electric connections. Integral safety controls are used along with electric or hydronic heat.

Water source heat pumps can be packaged units of either the console type or concealed horizontal or vertical type with ductwork. Cooling or heating functions are controlled either by manual switch or by a temperature controller which cycles the refrigerant compressor and controls a reversing valve to allow reverse cycle heat pump operation.

In the cooling cycle, space heat is absorbed on the roomside air evaporator coil. The space heat along with the compressor heat is rejected to a loop water cooled condenser coil.

In the heating cycle, heat is absorbed from the loop water by a water cooled evaporator coil and absorbed heat plus the compressor heat is rejected to space through a roomside air condenser coil.

The loop water temperature is allowed to vary from 60 to 95° F. When the loop water temperature drops below the minimum value, heat must be added to the system using a water heat exchanger.

When the loop water temperature rises to maximum value, heat must be rejected from the system by a closed circuit liquid cooler, cooling tower cooled heat exchanger, or directly by cooling tower. Since the loop water is circulated during winter, provisions must be made for the freezeup of loop water on the heat rejection side. Table 1-1 summarizes these different HVAC system classifications.

Table 1-1 HVAC System Classifications

All-Air Systems
> Single-path single-zone, draw-through
> Single-path multi-terminal
> Variable air volume (VAV)
> Induction reheat (IR)
> Constant volume reheat (CVR)
> Powered induction units (PIU)
> Parallel-path systems
> > • Multizone, blow-through
> > • Dual-duct, constant or variable volume

Air-Water Terminal Systems
> Fan-coil units, central ventilating air, 2, 3 or 4-pipe

All-Water Terminal Systems
> Fan-coil units with direct outside air intake, 2, 3 or 4-pipe
> Unit ventilators

Packaged Terminal Systems
> Ductless split
> Through the wall units with electric or hydronic heat
> Unit ventilators with direct expansion cooling and electric
> or hydronic heat
> Water source heat pump

Solar Heating Systems

There are four main parts to a solar heating system:

- The collector or absorber,
- The heat storage tank,
- Control system and
- Auxiliary energy source.

The sun warms the collector, which is normally covered by glass or other transparent material. The collector traps heat from the sunshine and this heat is transferred to a working fluid, such as water or water with an antifreeze mixture. The heated fluid is circulated through the collector and pumped, either directly or through a heat exchanger, into an insulated storage tank. When the water in the storage tank is hot, it can be used in the following ways:

- To supply hot water,
- As a heat source for a forced-air furnace, and
- To operate a heat-driven cooling system.

Auxiliary energy is used to supplement the solar system whenever the stored heat is used up. An auxiliary heating system using gas, electricity, or fuel oil can provide the needed heat.

A collector may be a flat metal plate insulated on its underside and painted black to absorb the heat of the sun. It may have passages through which the working fluid passes to be heated. These metal plates are made of aluminum or copper since these metals are good conductors of heat.

Insulation of the plate assembly minimizes heat losses. The assembly may be mounted in a frame with insulation on the back side and covered with glass. Since the unit is sealed, the glass traps solar radiation, which heats the insulated plates. This heat is transferred to the fluid flowing through the collector passages or tubes. This type of solar collector assembly is known as the flat-plate type.

Roof-mounted solar collectors are usually connected to a simple heat exchanger that is made up of several coils of tubing. When the heated fluid from the collector flows through these coils, it gives up or exchanges its heat to the water stored in the tank. The fluid is then pumped back to the solar collectors for reheating by the sun. This is a closed-loop system.

The fluid in the collector loop never comes into direct contact with the water in the storage tank. This type of system is used when there is a need to guard against corrosion, as with aluminum collectors, and where freezing weather occurs. In cold weather climates, an antifreeze mixture with corrosion inhibitors, similar to those used in automobiles, is sometimes used for the solar collector portion of the loop.

Solar Valves and Controls

There are several valves and controls used in a system of this type. When the collectors are hot enough to contribute heat to storage, controls operate the pump and circulate water through them. There is no pumping action on overcast days or at night since this would only cause heat to be lost to the atmosphere. The auxiliary heat only comes on when the temperature of the stored water is below the water heat thermostat setting.

For space heating or room heating, the heated water from storage is circulated through the coils of a heat exchanger located in the forced-air furnace. As the heated water flows through the coils, it radiates heat, which is distributed by the furnace fan and existing duct network. If this heat is not sufficient, the auxiliary heating element is used to boost the temperature.

Solar Cooling Systems

In a heat-driven cooling system, heat is provided by the gas flame. Solar cooling systems substitute hot water for the flame. Higher temperatures are needed for solar cooling than for water or space heating. If the water temperature in the storage tank reaches approximately 190°F, it can be channeled into the refrigeration unit, which provides chilled fluid for the heat exchanger. As the chilled fluid flows through the coils, a fan blows air through them. The air is cooled and flows through the distribution ducts.

Solar air conditioners of this type push the efficiency of flat-plate solar collectors. They may not provide the higher temperatures needed to effectively operate an absorption-type cooling unit. Marginal performance results if these higher temperatures are not reached.

Alternative Systems

The working fluid need not be a liquid, it can be air. In this case, the unit is referred to as an air system. Systems that use liquid heat transfer are called hydronic.

Most applications for an air system involve space heating or room heating, as opposed to air conditioning or water heating. As with the hydronic system, you still need a collector, a means of heat storage, a control system, and an auxiliary energy source. Since the warm air is used directly and will not freeze, no separate heat exchangers or antifreeze solution are needed. Rocks or gravel are typically used for heat storage in air systems.

In the air system, air flows over the metal absorber, picking up heat trapped by the glass cover. This heated air is then circulated through the air ducts if heat is needed. If heat is not needed, the air is diverted into the storage area where it heats the storage material. This stored heat may then be used when needed.

To retrieve the heat, air is blown through the rocks or other storage material, picking up the stored heat. An auxiliary heater warms the air as required by the setting on the thermostat.

Another technique is the passive system. This technique uses the building materials and construction design to store heat during the day and release it at night. Systems where fluid or air is forced to circulate by pumps or blowers are active. Passive systems use materials such as stones, bricks, adobe blocks, or concrete blocks to absorb heat and hold it for an extended period of time. They work best in desert-type climates, where the days are hot and the nights cold. Generally, they rely on the south wall of the house to collect and store heat.

Typical Systems

For space or room heating, both air systems and liquid systems are used. An air system with rock storage does not require a heat exchanger or any special precautions to prevent corrosion or freezing. It does have high pumping-power requirements. Adaptations of systems previously described are used in various combinations by different manufacturers.

Air Systems for Room Heating

Systems have been available from companies such as International Solarthermics as self-contained units, with the collectors, reflector, rock-storage unit, and blower fan in a single triangular-shaped container. This type of unit is designed to be mounted outside at ground level. The rock is stored inside the A-frame structure. The collector fan pulls air into the solar collector, through the stored rocks and a series of baffles and heat traps. The air is heated as it passes through a series of aluminum cups in staggered rows.

A differential thermostat is used. One sensor is in the solar collector and the other in the rock storage. The collector fan operates only when the temperature in the collector exceeds that of the stored rock by an amount large enough to increase the storage temperature. Another blower, called the distribution fan is also in the A-frame.

When the room thermostat calls for heat, the distribution fan on the solar furnace is turned on, forcing air through the heated rock. Cold air is drawn from the room back into the storage area to be reheated.

Liquid Systems

For room heating with a liquid medium, the solar collector piping is in a conventional closed-loop form, often circulating an antifreeze mixture. The operation of the solar heater is controlled by the room thermostat. If the inside air temperature falls below this point, heat will be supplied in one of two ways:

1. If the temperature of the solar-heated water in storage exceeds the minimum required for heating, the distribution pump starts and hot water from storage is circulated through the coils of a heat exchanger. The coils radiate heat, which is distributed by the blower and duct network.
2. If the storage temperature is below this minimum, an auxiliary electric heating element in the distribution duct supplies room heat. No pumping occurs.

Since solar energy is more cost-effective for applications that have electric heaters, the auxiliary heater can be powered by electricity.

A solar-heated water supply can be added to the system by connecting a conventional water heater in the hot water storage tank through a heat exchanger. It is possible to eliminate the exchanger in some systems, depending upon the type of storage tank and the composition of the water in storage. The quality of water entering the water heater must meet potable standards.

Some systems provide both space heating and hot water requirements. As the hot water is used, cold water can flow through a valve into the water heater or it can flow through the exchanger in the hot water storage, and then into the water heater. This can be done in the following ways:

- The temperature of the cold water is sensed by a thermostat. If it is less than the minimum, a three-way valve closes the direct flow, forcing water through the exchanger to absorb heat.
- When the water flowing through the exchanger is heated to a temperature in excess of the minimum, a sensor causes the valve to open the direct flow, mixing hot and cold water.
- Water, from either source, flows into the water heater. If the solar-heated water is insufficient to maintain the minimum temperature in the water heater, the thermostat activates the electric resistance heater, boosting the temperature.

Solar Plus Heat Pump

Adding a heat pump to the system provides solar-assisted room heating plus hot water from central storage. In addition, the heat pump provides air conditioning, but it does so without any contribution from the solar system.

The heat pump acts like an air conditioner with reversing valves. When the fluid flows in one direction, heat is produced. When it flows in the opposite direction, heat is absorbed resulting in air conditioning.

The heat pump operates like a conventional vapor-compression type home air conditioner, except it can heat, as well as cool, by reversing the cycle. Like an air conditioner, a portion of the heat pump is exposed to the outdoors. When the heat pump is operating as an air conditioner, it absorbs heat inside the house and ejects it outdoors. It does this operating on electrical power, with no help from the solar system. For the heating mode, its operation is reversed. The pump absorbs low-grade energy from the outside air, warms it to the desired temperature, and circulates it inside the house. Here, solar energy can assist the heat pump, providing higher temperatures from water storage than are available from the outdoors. This increases the efficiency of the heat pump and lowers the amount of electrical energy it uses.

Most heat pumps are not effective when outdoor temperatures are below 30°F. Without solar storage, the electric resistance heater in the pump package must provide all of the heating at a very high cost. On the coldest days in winter, which tend to be clear, the solar collectors operate at their best efficiency, the heat pump at its lowest.

In a solar-assisted heat pump these devices combine into a more efficient heating system. Both units can then be smaller in size than if used individually, resulting in cost savings. Although individual system designs can vary, a typical system might function as follows:

- If the water in the storage tank is at a minimum useful temperature or greater, it is circulated directly through the heat exchanger bypassing the heat pump.

- If the stored water is at a temperature between 40°F and 103°F, and above the outdoor temperature, the stored water becomes the heat source for the heat pump.
- If the outdoor temperature is above that of the stored water, it then becomes the heat source for the heat pump.
- If the heat pump is operating, and it is unable to meet the room temperature setting of the thermostat, an auxiliary electric-resistance heating element is turned on.

Absorption Cooling

Solar-assisted air conditioning can be accomplished using absorption refrigeration machines. Refrigerators that operated from a gas flame were made by the Servel Company, which later became the Arkansas-Louisiana Gas Company (Arkla). They have also manufactured gas-fired air conditioners that were converted to use solar heat instead of gas. This type of absorption machine uses lithium bromide to absorb the refrigerant vapor, changing it to liquid. Changing the refrigerant from vapor to liquid and back to vapor again is the basic cycle that causes cooling. Conventional air conditioners or refrigerators accomplish this using a mechanical compressor.

The solar-heated water from storage replaces natural gas as a heat supply to the unit, and much higher water temperatures are required to operate the air conditioning unit efficiently. A hot water inlet temperature of 210°F may be used. Only the most efficient flat-plate collectors using dual glazing and selective surfaces are suitable for these applications.

There are times when the solar-heated water in the storage tank will be below the minimum temperature necessary to operate the absorption air conditioner. When this occurs, an internal heater is used to boost the water temperature.

Air conditioning is also possible using the Rankine solar cooling system. Solar-heated water at 200°F is used to vaporize a refrigerant such as Freon. This drives a turbine to operate the compressor in the heat pump. This system can be used to cool with electricity on a hot, overcast day.

Expansion Tanks

An expansion or surge tank may be used on the outlet side of the collector. The expansion tank allows for any boiling of the fluid in the collector or pressure buildup that could be caused by a power failure. The expansion tank allows for fluid expansion, but it also keeps the collector loop filled with fluid.

Differential Thermostats

A solar temperature controller or thermostat must measure and compare two temperatures. A typical unit has two heat sensors, one sensor is attached to the surface of a solar-collector absorber plate and the other at the water storage tank. When the temperature of the collector exceeds that of the stored water by a specified amount, a relay in the unit closes, starting the collector pump motor. The pump circulates the fluid through the collector loop of the system. The thermostat prevents pumping action on overcast days or at night. The differential thermostat should prevent unnecessary on-off cycling of the pump. This is done with the temperature differential circuit. The typical operating cycle for one day's operation is described.

The collector pump is initially off. As the sun rises, the collector temperature will increase above the storage tank temperature. At about 20°F above the tank temperature, the thermostat relay closes, starting the pump motor. As the collector fluid in the tank begins to circulate, it causes an initial drop in the collector temperature. As long as the collector temperature remains several degrees above the tank temperature, the pump continues to operate.

As evening approaches, the collector temperature falls. When it is less than a few degrees above the tank temperature, the pump is turned off. Once fluid flow stops, there is a temporary rise in temperature, but as long as the temperature at this point is less than about 20°F above the temperature of the tank, the pump remains off until sunrise. This type of differential-thermostat control system assures that maximum heat energy is stored and retained in the tank under various weather conditions.

Some units have the solar collector sensor in the waterline, where the water leaves the last collector, instead of attaching it to the absorber plate. This directly measures the temperature of the water leaving the collector.

For systems that use airflow rather than a liquid flow, the sensor is attached directly to the absorber. The sensor that is used to measure the storage temperature may be located either inside the tank near the bottom or in the exit waterline leading to the pump. If externally located, it should be as close to the tank-end of the line as possible.

Some differential thermostats also have a circuit that closes a relay when the collector-sensor temperature approaches the freezing point of the collector fluid. In an area where freezing temperatures occur and an antifreeze solution is not used in the collector fluid, this feature can prevent collector damage due to freezing. There are two functions that may be controlled by this relay:

- Opening valves to drain the collector and
- Activating electrical heating elements in the collector to warm them above the freezing point.

Variable Flow Control

Another control that may be incorporated in the differential thermostat is variable flow control. This involves the control of the speed of the circulating pump in the collector loop as a function of the absorbed temperature. A slower flow of liquid through the absorber results in higher outlet temperatures than a faster flow. There is an optimum balance, since the collector efficiency decreases as the absorbed temperature increases due to heat losses. Varying the flow rate will help to improve the overall efficiency of the system.

A proportional (variable flow) controller is used to control the flow rate. If the pump has too high a flow rate for the solar heating level, the pump will go from on to off once fluid flow starts and the absorber is cooled. As the pump turns off, the flow goes back to zero and the absorbed temperature climbs again. This cycling of the pump and unstable operation will continue until the solar heating level increases and the higher flow rate can be sustained. Flow control will allow a smooth, continuous operation along with greater efficiency.

A proportional controller can activate a switch on the motor, changing its speed at the correct level. Some circulating pumps use shaded-pole motors, where the speed can be varied over a wide range. Others use a permanent split-capacitor type for two speeds only. The split-capacitor type provides better electrical efficiency and a higher starting torque for pumping water against a head.

Control Circuits

Control circuits in solar heating systems typically set the temperature of the storage tank Ts plus a user-designated temperature differential Td to the temperature of the collectors Tc.

$$Tc = Ts + Td$$

When the controller finds that this equation is satisfied or exceeded, it turns on a pump or blower to activate the heating system. In a typical system, several pairs of temperature sensors might have to be used.

In many circuits a separate amplifier is used for each temperature sensor. For differential measurements, the outputs of two of these amplifiers are fed to another amplifier. Another type of circuit uses multiplexers to handle the signals from temperature sensors. The circuit uses electronic-switch arrays, synchronized by a common oscillator and logic circuit, to multiplex the signals from the temperature sensors. In a typical system the sensors would be thermocouples or thermistors. The switches connect the channels in the same numerical order so that at any instant the same channel is routed through all the multiplexers.

Solar Controllers

Controls for solar heating generally consist of a differential controller and a temperature monitoring system for the solar-powered heating components. A typical solid-state controller is a differential thermostat that monitors the temperature in the collector and in the hot water storage tank of the solar heating system. It switches on when the collector temperature exceeds the storage temperature by a preset limit, typically 20°F (11.1°C). In the on condition the input voltage, either 120 or 24 Vac, is delivered at output terminals for the control of pumps in the collector loop.

These controllers are used with either air or water temperature probes containing matched sensors in metal housings. The probes are electrically connected to the controller package through high-temperature-resistant cables. Typical tracking accuracy is +/-5°F (+/-2.7°C) over the operating temperature range between -40 and +300°F (-40 and + 149°C).

The temperature monitor, which may also be a solid-state device, senses the indoor, outdoor, and storage temperatures. Linear diodes that operate over the range -40 to +199°F (-40 to +111°C) are often used as temperature sensors. A thermostat generates signals to control the solar and backup heating systems. Light-emitting diodes may also indicate the mode of operation as solar collecting, solar distribution, or backup heating.

Solar Heating Modules

Typical of the solar heating modules in use are those developed by the Johnson Environmental and Energy Center of the University of Alabama in Huntsville (UAH). The prototype unit was installed at the Alabama Space and Rocket Center in Huntsville and used to heat the Space Odyssey, a simulated space ride at the center.

A solar collector array of 240 square feet is used. About 500 gallons of hot water storage is provided as a source of wash water for tour busses and mobile exhibit vans stationed at the Alabama Space and Rocket Center, in addition to meeting the space heating requirements for the space ride building.

The module, which is housed in a small building with the collectors mounted on the roof, uses water as a transfer medium. The heated water is pumped to a storage tank and can be transferred from the tank through an airhandler to provide heated air.

Solar Controls and Components

The solar controls use a differential thermostat to sense the difference in the collector array temperature and storage tank temperature. A collector array sensor may also be used to sense near freezing temperatures as a means of initiating shutdown and drain down. The drain down valve is normally open and provides the solar collectors to be in a drain down condition when there is insufficient solar energy, power failures, or near freezing conditions. A three-way valve is used to assure that the coldest transfer fluid is at the collector array inlet manifold.

A tank temperature thermostat provides a method of disabling the heat demand signal when the storage tank temperature is below 100°F. This assures that there is sufficient stored energy for proper operation of the system.

Solar System Components

The airhandler motor is a 208VAC 3-phase unit which provides heated air on demand. The pump motor is also a 208VAC, 3-phase unit. It pumps liquid from the storage tank through the coils. The collector pump motor is a 208VAC, 3-phase unit which pumps transfer fluid through collector array. Separate magnetic motor starters are used as slaved relays. These relays allow on/off control from a remote signal source.

Two double-pole single throw (DPST) 120VAC relays are slaved to the differential thermostat output and provide additional switching capability to the drain down valve and collector pump motor. Two other DPST 24VAC relays are slaved to the heat demand thermostat. These provide control to the airhandler pump motor and airhandler blower motor.

Modes of Operation

There are three modes of operation:

- No demand mode – Solar energy is available, but there is no demand for the use of energy in the area to be heated.
- Energy demand mode – Solar energy is available and there is a demand for energy in the area to be heated.
- No energy mode – No solar energy is available, but there is a demand for energy in the area to be heated.

Chapter 2
Optimizing HVAC

The airhandler is the basic unit of space conditioning. It is used to keep occupied spaces comfortable (Figure 2-1) or unoccupied spaces at desired levels of temperature and humidity. In addition to supplying or removing heat and/or humidity from the conditioned space, the airhandler can also provide ventilation and fresh air movement.

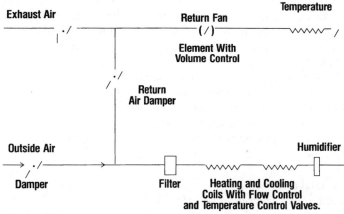

Figure 2-1 Airhandler

Depending on the use of the space involved, from 75,000 to 300,000 BTU/year (19,000 to 76,000 calendar/year) are required to condition one square foot (0.0992 square meters) of building space. The cost of the energy used varies, this produces a yearly operating cost of several dollars per square foot of space.

Other unit operations have improved substantially from advances in process control, however, airhandlers have not. Airhandlers are often controlled the same way as they were decades ago. This allows airhandler optimization gains of much greater percentages of savings compared to the optimization of other plant operations.

Optimization can often reduce the cost of airhandler operation by 50%. This type of savings is extremely difficult to achieve in most types of system operations.

Typical optimization goals and strategies include the following:
- Self heating lets the building heat itself,
- Takes advantage of any available free cooling and/or free drying,
- Utilizes gap control or zero energy band (ZEB),
- Eliminates or reduces losses due to chimney effects,
- Optimizes the startup timing to minimize energy use,
- Optimizes the air makeup, CO_2 levels, makes maximum use of the supply air temperature,

- Minimizes the fan energy,
- Automate/maximize the operating modes,
- Minimize reheating, and
- Automate/maximize the balancing of air distribution.

Airhandler Operation

Heating, ventilation, and air conditioning controls can optimize comfort in laboratories, clean rooms, warehouses, offices, and manufacturing areas. Supply air is the means of providing comfort in the conditioned zone. The air supplied to each zone provides heating or cooling, adjusts the humidity, and refreshes the air. The control system must be able to optimize the temperature, humidity, and fresh air ratio in the supply air. The components of the airhandler include fans, heat exchanger coils, dampers, ducts, and instruments. The following operations may take place:

1. Outside air is admitted by an outside air damper.
2. This air is then mixed with the return air from a return air damper.
3. The mixed air is filtered, heated or cooled and humidified or dehumidified as required by the control system.
4. The resulting supply air is then sent to the conditioned zones by variable rate supply fans.

Zone Control

In each zone a variable air volume damper can determine the amount of air required and a reheat coil adjusts the air temperature as needed. The return air from these zones is transported by variable-rate, return-air fans. When the amount of available return exceeds the demand for it, the excess air is exhausted by an exhaust air damper. The conditioned spaces are normally pressurized to about 0.1 inches H_2O (25 Pa), relative to the barometric pressure outside. This pressurization will cause some air leakage through the walls and windows, which depends on the type and quality of the building construction. The outside air must balance the exhaust air plus any pressurization loss. In the normal mode of operation, the airhandler will operate with about 10% outside air. In the purge or free cooling modes, the return air damper is closed, the outside air damper is at full open, and the airhandler operates with 100% of outside air.

HVAC Process

If we consider HVAC as a process, the process material is clean air, the utility is hot or cold water or steam. The overall system behavior tends to be slow, stable, and forgiving so proportional control is often used with low-quality instruments and control. These are even operated in a basic on-off fashion with a very basic control loop.

The advantage in applying higher quality process control equipment and control loops to the HVAC process allows major reductions in operating costs due to the increased efficiency of operation. One of the most efficient control concepts is based on the operating mode selection. The correct identification and timing of the various operating modes can add much to the overall optimization of the building. Typical operating modes include:

- Startup,
- Occupied,
- Unoccupied or night and
- Purge or free-cooling.

Reducing the time of startup lowers the cost required in getting a building ready for occupancy. One approach is to calculate the amount of heat that needs to be transferred and divide it up according to the capacity of the startup equipment.

A computer control system can initiate the unoccupied or night mode of operation. It can also recognize weekends and holidays and provide a flexible means for time-of-day controls.

Free Cooling

The purge or free cooling mode is another useful tool for optimization. When the outside air is preferred to the return air, the building is automatically purged with fresh air. This free cooling can be used on dry summer mornings or free heating can be provided on warm winter afternoons. Purging is like opening the windows.

In computer-optimized buildings, an additional gain is to use the building structure as a means of heat or coolant storage. The purge mode can be initiated during cool nights prior to hot summer days. This will bring the building temperature down and store some free cooling in the building structure.

Summer/Winter Mode Switching

Another important mode feature is switching from summer to winter mode and back again. Conventional systems are switched according to the calendar, while optimized systems can recognize there may be summerlike days in the winter and winterlike hours during summer days. Seasonal mode switching can be wasteful.

Optimized building control can be provided by making the summer/winter switch based on the enthalpy. When heat needs to be added, switch to winter mode and if heat needs to be removed, switch to summer mode, regardless of the actual calendar.

In airhandlers that serve a variety of zones, it is important to first determine if the unit is in a net cooling or summer mode or a net heating or winter mode before the control system can decide if free cooling or free heating by the outside air should be used.

Figure 2-2 shows the basic heat balance control loop that can be used to determine the mode of operation. The heat balance calculation should be done every 15 to 30-minutes and can easily be implemented through the use of computers.

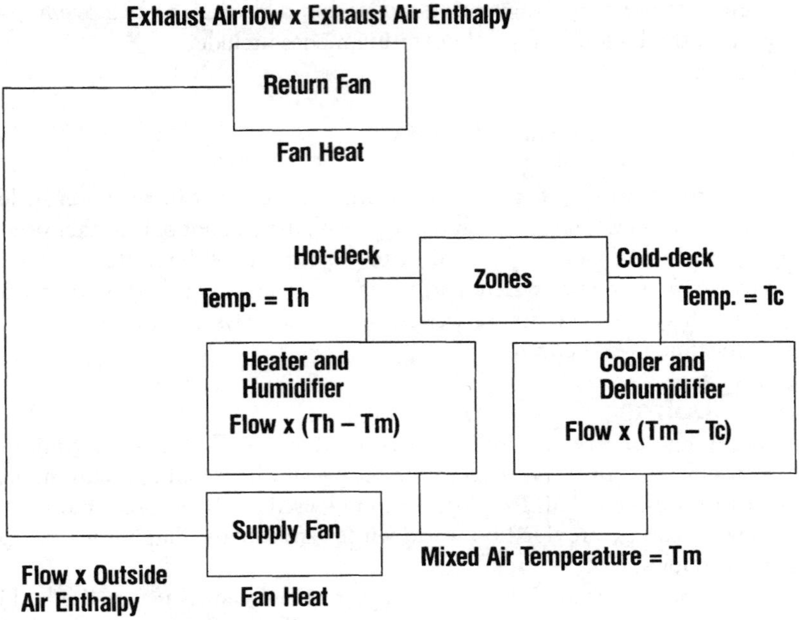

Exhaust Airflow x Exhaust Air Enthalpy

Figure 2-2 Summer/Winter mode determination

Emergency Mode

Another useful feature allows the airhandler to have an emergency mode. If fire, smoke, freezing temperature, or pressure conditions exist, the emergency mode can keep these conditions from spreading.

Table 2-1 shows the status of fans, dampers, and valves in each of the operating modes. In a computer control system, the mode selection and the setting of the actuated devices are programmed automatically.

When a smoke or fire condition is detected in the return or supply ducts, the fans stop, the outside air and return air dampers close, the exhaust air damper opens, and an alarm is actuated. The airhandler can be switched into the purge mode, with the fans started, the outside air and exhaust air dampers opened, and the return air damper closed.

If a smoke/fire emergency requires fire fighters, a fire command panel will allow the fire chief to operate the fans and dampers as needed for a safe and orderly evacuation and protection of the building.

Freeze Protection

Another emergency condition occurs when a freezestat switch on one of the water coils is actuated. These switches are normally set at about 35°F (1.5°C). They are used to protect coil damage resulting from freeze-ups in systems using water coils, either chilled or heated water. The capability to introduce large percentages of outside air requires that a low limit temperature controller or freezestat be provided.

Table 2-1 Mode Control and Alarms

	SF	RF	OAD	EAD	RAD	CCV
Mode						
Off	–	–	C	C	O	C
On	Off	Off	<------Modulating------>			
Warm-up	Off	Off	C	C	O	O(HC)
Cool-down	Off	Off	C	C	O	O(CC)
Night	< Cycled to maintain required nighttime temperature >					
Purge	On	On	O	O	C	Modulating
Alarms						
PSH,RFOUT		Off		C		
PSL,RFIN		Off		C		
S/F,RA	Off	Off	C	O	C	C
TSL,COILS	Off		C		O	C
PSL,SFIN	Off		C		O	
PSH,SFOUT	Off		C		O	
S/F,SA	Off	Off	C	O	C	C
Key:						

Key:

SF	= Supply Fan	RAD = Return Air Damper
RF	= Return Fan	CCV = Coil Control Valves
OAD	= Outside Air Damper	PSL = Pressure Switch Low
EAD	= Exhaust Air Damper	PSH = Pressure Switch High
TSL	= Temperature Switch Low, Freezestat	

The low limit controller can be a modulating-type temperature controller. The averaging bulb element is installed in air leaving the cooling coil. The outside air quantity can be controlled as required to raise the mixed air temperature above the freezing point. This may require opening the fan shutdown loop circuit to stop the supply fan.

A single stage freezestat system will shutdown the fan, close the outside air damper, and activate an alarm. A multistage freezestat system may operate with the following stages. At 38°F (3°C) the outside air damper is closed. At 36°F (2°C) the water valve is opened and at 35°F (1.5°C) the fan will be shutoff.

It is important that the sensing element be located in the coldest part of the mixed air. This must be done by actual temperature measurements during sub-freezing weather.

If both face and bypass damper control is used on a preheat coil, stratification can result if the hot and cold airstreams do not mix downstream. A sub-freezing cold airstream can cause nuisance switching of the freezestat. To avoid this, a mixing baffle can be added in air leaving the coil face and bypass dampers. This baffle should mix the two airstreams to a temperature above the freezestat setpoint.

The control functions for air moving systems include start-stop, emergency shutdown, smoke damper operation, smoke removal, and outside air control.

Start-Stop

The on-off or start-stop sequence of an airhandling unit may be controlled in the following ways:

- Manually with a starting switch or hand-wound interval time,
- Automatically through a programmed timing device,
- Automatically from a direct digital control (DDC) system, or energy management system (EMS), or energy management and control system (EMCS).

The manual starting switch method is often used only on a system which runs 24-hours a day. For spaces with daily or weekly usage programs, an automated start-stop sequence is used. The power to fans may be energized through auxiliary contacts on the supply fan starter, relays on the load side wiring of the supply fan motor starter, or by an airflow switch in the supply duct.

Emergency Shutdowns

Provisions for emergency fan shutdown are required to comply with the requirements of fire codes for life safety and for safety of the contents and the structure. The fire code most often used in the United States is NFPA 90A, Installation of Air Conditioning and Ventilating Systems. The code requires a manual emergency stop means for each air distribution system to stop the operation of supply, return, and exhaust fans in event of an emergency. Smoke detectors must also be located downstream of air filters and upstream of any supply air takeoffs in all systems larger than 2,000 cfm capacity.

Smoke detectors must also be installed in systems larger than 15,000 cfm capacity serving more than one floor at each story prior to the connection to a common return and prior to any circulation or outside air connection.

The manual emergency shutdown is intended to stop the fans in order to prevent spreading fire and smoke through the duct system before the automatic smoke or temperature-based shutdown devices turn on. An emergency stop switch must be located in an area approved by the authority having jurisdiction, which is usually the local fire marshal.

For small systems the electrical disconnect switch can be used for the emergency stop switch. In larger systems a separate break-glass station or connection with the fire alarm system provides the means for an emergency shutdown.

Firestats

Older buildings usually have manual reset fixed temperature devices, called fire safety thermostats (FST) or firestats. These are found in systems from 2,000 to 15,000 cfm and were installed to comply with fire codes which were current at time of construction for automatic shutdown from high temperatures.

These firestats open when the temperature sensed is above the setting. Firestats are wired to interrupt the holding coil circuit on the magnetic starter of the primary fan or fans.

Firestats mounted in return airstreams are usually set at 125°F to 135°F. Firestats in supply ducts are often selected with setpoints similar to fire dampers. These setpoints will be based on the normal temperature to be transmitted in the duct on the heating cycle and are set no more than 50°F higher than that temperature.

Smoke Detectors

Large systems in older buildings and systems larger than 2,000 cfm capacity in newer buildings may have smoke detectors installed instead of firestats. The most common smoke detector for duct systems is the self-contained ionization type.

When smoke detectors are used in a building having an approved protective signalling system, the activation of the smoke detector in the air distribution system must cause a supervisory signal to be indicated in a constantly attended location or initiate an alarm signal.

When smoke detectors are installed in a building that does not have an approved protective signalling system, the activation of the smoke detector must cause an audible and visual signal to be indicated in a constantly attended location. Trouble conditions in the smoke detector must also be indicated either audibly or visually in a normally occupied location and identified as a duct detector trouble condition.

Other fire and smoke detection devices can be wired into an emergency fan shutdown loop. These include alarm contacts in a manual fire alarm system and sprinkler flow contacts in an automatic fire protection sprinkler system.

Start-stop control of ventilating and exhaust fans may be directly controlled from the building automation system, interlocked with the supply fan, or controlled from a temperature controller, thermostat, other ambient condition sensing device, or hand-wound interval timer.

Another type of emergency occurs when there is excessive pressures in the ductwork on the suction or discharge sides of the fans. This can result from operation against a fail closed damper or other equipment failure. When this happens, the associated fan is shutdown and an alarm is actuated.

Terminal Devices

Terminal devices are the devices used to provide the final control element for the conditioned space. They include: zoning dampers, constant volume boxes, variable volume boxes and powered mixing boxes.

Zoning dampers, which are also called zone mixing dampers, are found where the end of a duct run is under the control of a separate zone damper. Suppose a small eastern exposure zone is served on the end of a southern exposure zone. The zoning damper can reduce the airflow to the eastern zone as the sun load drops off.

The control of zoning dampers is usually done via controller to damper actuator. The output from zone temperature controllers can be used for the reset of cold-deck and hot-deck temperatures.

Electric Relays

Electric control relays are usually the enclosed dry contact type with silver-plated contacts rated for 5 amperes at 240 volts. The available relay actions are single-pole-single-throw (spst), double-pole-double-throw (dpdt), or multipole, multiple-throw.

Delay relays are available with timed-opening or timed-closing contacts. Electric-pneumatic relays are solenoid air valves with normally open, normally closed, and common ports. These are used to connect and disconnect the air supply to a system and exhaust the pressure to atmosphere.

Transducers

These are devices which convert a signal of one type to a proportional signal of another type. Transducers in HVAC control may need to interface with pneumatic systems. In a modulating electronic to pneumatic transducer, the signal, which may range from 1 to 15 volts, is converted to a proportional pneumatic signal. This can also result in an on or off switching of the system air pressure range.

A pneumatic signal of 3 to 15 psig, or other range, can be converted to an electric resistance. Typically, this is 135 ohms in electrical systems and 1,000 ohms in electronic systems. The pneumatic signal may also be converted to a voltage. Typical voltages are 1 to 5 vdc or 1 to 15 vdc. Duct pressures as low as 0.0005-inches of water can be converted to a 4-20 mA, 0-5 or 0-10 vdc voltage output. Damper position indicators include potentiometer devices that are connected by a linkage to the damper actuating mechanism. These provide an analog output proportional to the damper position.

Motor Starting

Electric relays are used in the starting circuits of magnetic starters and for on-off controls of other items. An input signal from the building automation control system will start/stop the motor or energize/deenergize the controls. Most of this equipment will have safety controls.

Enthalpy and Temperature Measurement

The measurement of enthalpy or total heat content of air can be difficult to perform accurately and repeatedly. Older systems have used wetted sleeves on temperature sensors to give the wet bulb temperature readings. More recently, a number of relative humidity sensors have been developed which operate on the following principles

- Measuring the capacitance of a thin film element,
- Measuring the electrical resistance of a salt solution,
- Measuring the resistance of carbon particles in a hygroscopic envelope,
- Measuring the opacity of a chilled mirror for dew point readings.

Dry-bulb temperature measurements can be made with conventional electronic sensors such as resistance temperature detectors (RTDs), Balco resistance elements, or thermistors. The first two devices are stable over long periods and are preferred.

Damper Position

A feedback signal from the dampers can come from a damper position indicator that is attached to the damper linkage. The proportional signal is an electrical resistance value which is calibrated to provide the damper position indication.

Hot-deck and Cold-deck

The terms hot-deck and cold-deck refer to the heating and cooling sections of a multizone or dual-duct system. The term hot-deck can be misleading since the action of the cold-deck and hot-deck temperature controls is often intended to fade-out the hot-deck as the outdoor temperature rises so that the hot-deck becomes a bypass deck during the cooling season. In parallel path systems, where one airhandling unit is serving several zones, a separate temperature controller for each deck maintains the deck supply air temperature. Chilled water coil supplied cold-decks on systems which do not operate 24-hours per day may not be temperature controlled.

Multi-zone mixing dampers often use two damper blades mounted on a common shaft. The position of the cold-deck damper is 90° degrees from the hot-deck damper. When one section is 80% open, the other will be 20% open. The flow rate through each mixing damper segment is greater than its percentage opening. For example, at 20% open the airflow may be 90% of fully-open flow. The total flow through a damper can be greater than the percentage opening for the two damper segments. This causes some variation of airflow among zones as the damper positions are changed and the fan shifts its operating point on the fan curve.

CVR Boxes

Constant volume reheat (CVR) terminal boxes are single-inlet boxes with either heated water or electric resistance reheat coils. Constant volume regulators tend to be mechanical devices built into the box. They generally have manual setpoint adjustments that are made during system commissioning, as a part of the testing and balancing phase. The only automatic control function is the temperature control of the reheat coil.

CVDD Boxes

Constant volume dual-duct (CVDD) terminal boxes are used with cold-air and hot-air ducts which are temperature controlled using cold-deck and hot-deck temperature controls. Constant volume regulators are mechanical devices built into the box.

Temperature control functions vary between terminal box manufacturers. The basic method is to control the cold-air volume with a variable volume valve at the box inlet and to control the hot-air volume using a constant volume regulator.

When the cold-air volume required to meet the cooling loads is less than the constant volume setting, the difference between the required cold-air volume and the constant volume setting is made up by hot-air. When the required cold-airflow to meet the space cooling load equals the constant volume setting, no hot-air is allowed to enter the box.

The constant volume regulator can be pressure dependent, but some boxes use airflow sensors instead of pressure sensors. Pressure dependent regulators allow the delivered air volume to change in response to input pressure changes. Pressure independent regulators hold the delivered air volume to a selected value over a large range of input pressures.

VAV Boxes

Variable air volume (VAV) boxes or valves and variable volume reheat (VVR) boxes use a single duct inlet. VVR boxes are like VAV boxes or valves with a reheat coil. The coil may use either heated water or electric resistance. Some VAV boxes or valves are pressure dependent and used with only cfm limiting applications. Others are pressure-independent and act as variable constant volume devices. The setting of their constant volume regulator is controlled from the space temperature.

VAV boxes can have minimum airflow cfm settings at a level which keeps the airflow from decreasing below acceptable level of air motion in the space. The airflow can also be allowed to close off completely to prevent any overcooling of the space.

VVR boxes can have dual minimum airflow cfm settings which provide about a 40% to 60% minimum setting for adequate airflow over the reheat coil. The dual minimum settings are indexed by the central control system in a similar way to summer/winter changeover. There is also a complete shutoff during the cooling cycle to minimize overcooling.

Control of VAV boxes is done from a space temperature controller which controls the volume regulating device over the full throttling range. Control of

VVR boxes is done with a space temperature controller which controls the volume regulating device down to a minimum position in the upper half of the throttling range and then it positions the reheat coil control in the lower half of the throttling range. For 100% shutoff boxes, the control for the zero minimum is the same as for a VAV box during the cooling cycle.

Controller Diffusers

These devices also provide variable volume performance. The most widely used type is self-actuating and uses a beeswax-filled cylinder in the airstream. This cylinder responds to space temperature changes by expanding or contracting. This actuates a linkage which move damper blades in the diffuser body and varies the volume of air delivered. The controllable diffuser can have a bimetal strip changeover switch in the supply air to shift the action of the damper blades for cooling or heating.

Balancing

The balancing of terminal units for airflow depends on the specific type of terminal device. Zoning dampers and multi-outlet terminal units are tested with the sum-of-the-outlets method, where all the outlets served by a specific terminal device are measured for airflow and proportionally balanced. The total airflow is then balanced by positioning the zone balancing damper or by adjusting the air volume regulator setpoint.

Terminal devices serving a single outlet are often balanced directly by testing the outlet flow and then positioning the zone balancing damper, or by adjusting the air volume regulator setpoint. The balancing of outlets should be done by adjusting the branch damper. The damper at the outlet should only be used for changing the air pattern, so that the increased noise generated in increasing the pressure loss for balancing will be attenuated. Enough airflow should be maintained across electric resistance heaters to prevent nuisance tripouts and possible element burnouts caused by overheating.

Fan Control

In a variable volume system the supply air volume delivered varies with the building load and it is necessary to provide some means for fan capacity modulation and static pressure control in the duct system. Static pressure control will prevent excessive noise generation and energy loss as the variable air volume system reduces the volume of air delivered. The techniques used include riding the fan curve, return air dumping, discharge damper control, variable inlet vane control and fan speed control.

The first technique depends on built-in fan performance. The second controls air direction without a reduction of volume and the others modify fan performance. The fans in HVAC systems are usually centrifugal blowers. They may have forward-curved, backward-curved or airfoil-bladed wheels. Each type has a different pressure-volume-horsepower relationship but the performance of all types can be predicted based on the air volume handled, the static pressure developed, and the horsepower required to maintain the fan speed.

The forward-curved wheel has a characteristic curve which shows that horsepower increases and decreases in direct relation to the volume of air delivered when operated at a constant speed. The backward-curved and airfoil-bladed wheels have a limiting value at a given speed beyond which no further pressure will be developed or horsepower will be required. But, the power requirements are not reduced in proportion to volume reductions as closely as forward-curved bladed wheels.

The characteristics of forward-curved bladed wheels allow fan modulation by using or riding the fan curve. No action is taken to modulate the fan speed and the fan is allowed to shift its operating point on its characteristic curve as the pressure and volume changes.

When the fan is operating at a constant speed, as a variable volume air damper acts to decrease the air volume delivered, the pressure developed by the fan will increase to a new operating point on the constant rpm curve and the power required will usually decrease.

When using the fan curve, the air volume delivered is varied by the terminal unit dampers with no control at the fan. Some terminal units are not able to control the air volume at high inlet pressures that result from using this technique. Some other static pressure control method must be used to prevent overpowering the terminal boxes and affecting the temperature control.

Although this technique can be used with other types of fans, the reduction in power required will not be as great as with a forward-curved bladed wheel. This usually makes other fan modulation systems more economically attractive.

With return air dumping, the pressure is sensed by the duct static pressure controller. When it increases above the setpoint of the controller, an actuator opens a bypass damper installed between the supply and the return air duct.

When the air volume delivered by a fan is to be reduced, the static pressure can be increased by damping down the air supply from the fan discharge. An automatic static pressure control damper of this type is called a discharge damper. A static pressure controller or a controller with a remote static pressure sensor is generally used.

The performance characteristics of any fan can be altered by changing the speed, or by changing the airflow direction at the inlet to the wheel. When a radial blade damper is placed at the inlet to the fan, with a control actuator connected to change the angle at which the radial blades cut the air entering the wheel, the blades are known as variable inlet vanes. They can be used to modulate all types of centrifugal fans but they are most efficient with airfoil-bladed and backward-curved fans.

Some typical fan controls are shown in Figure 2-3. Each zone is supplied with air through a thermostat controlled damper which is a variable air volume box.

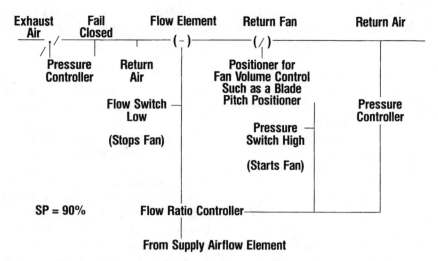

Figure 2-3 Variable-volume fan control for return air

The VAV box openings in the different zones determine the total demand for supply air. The pressure in the supply air distribution duct is monitored by a pressure controller which modulates the supply air fan to match the demand.

When the fan capacity is pushed to its maximum, a pressure switch starts an additional fan. When the demand for supply air drops, a flow control switch stops one fan unit if the load can be met by fewer fans.

The fan units are started on pressure and stopped on flow control. The operating cost 20 to 40% lower compared to constant-volume fans with conventional controls. Using variable-speed fans can also save significant amounts of energy.

Since the conditioned zones have a slight overpressure, some of the conditioned air leaks out creating some pressurization loss. Controlling this pressurization loss is done using the following scheme.

A flow ratio controller between the exhaust and supply air systems is set at 90%. This allows the return-air fan to return 90% of the air supplied to the zones and controls the pressurization loss at 10%. which meets the minimum fresh-air requirement and results in minimum-cost operation.

Since the conditioned zones may represent a relatively large volume, a change in the supply airflow will not immediately result in a need for a corresponding change in the return airflow. A pressure control is used in the system to prevent the flow-ratio controller from increasing the return airflow rate faster than required. This balancing reduces cycling and provides protection from collapsing the ductwork from excessive vacuum.

When the control system is properly tuned and balanced, the outside air admitted into the airhandler will match the pressurization loss, and no return air will be exhausted.

The dampers should also have a tight shutoff. It is cost-effective to install tight shutoff dampers because the resulting saving over the life of the building is much greater than the increased cost of the dampers.

Damper Pressure Drop

If dampers are to provide good control, a reasonable amount of the total pressure drop should be assigned to them. This is usually a differential pressure of about 10% of the total system drop.

Excessive damper drops should be avoided, since they increase the operating costs of the fans. Outside and return air fans should be sized for 1500 fpm (457 m/min) velocity at maximum flow.

Airstreams may be mixed or ratioed using outside and return air dampers. The damper differential pressure should be relatively constant as the ratio is varied. Parallel blade dampers are more efficient in this application and give superior performance. Opposed blade dampers develop too much pressure drop at the 45% opening and these dampers will cause the airhandler to be starved for air whenever the ratio goes through 50-50%. Parallel blade dampers exhibit a minimum pressure drop at 45%.

Fan Energy

The typical airhandler uses more fan energy than necessary. This is because the return air fan is sized to generate the pressure needed to exhaust the air from the building. This means that the pressure drop across the return air damper is about three times greater than necessary. Usually the return fan is located before the return air damper (Figure 2-1). This places the return fan and supply fan in series, they are essentially moving the same air, except for building losses. A more efficient configuration is to move the return fan outside this loop as shown in Figure 2-4.

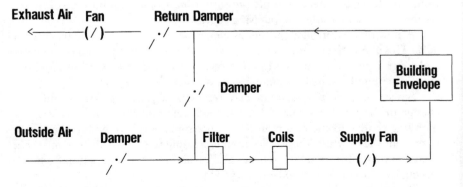

Fig. 2-4 Improved airhandler configuration

This configuration eliminates the waste of fan energy. Since only the sup-ply fan operates continuously, this reduces the pressure drop across the return air damper.

The return fan is started only when air needs to be changed, and its speed is varied to adjust the amount of air to be exhausted. Relocating the return fan also removes its heat input, which in most systems represents an added load on the cooling coil.

Temperature Control

Space temperatures are usually controlled by thermostats. However, the thermostat is a proportional-only device. The output signal is close to a straight-line function of the temperature, according to the following relationship:

$$T = K_C (T_M - T_N) + T_O$$

where
T = output signal
K_C = proportional sensitivity (K_C can be fixed or adjustable)
T_M = measured temperature
T_N = normal value of measurement, corresponding to center of the throttling range
T_O = normal value of the output signal, corresponding to the center of the throttling range of the control valve or damper

The term used to describe the sensitivity of thermostats is throttling range. This term refers to the amount of temperature change that is required to change the thermostat output from its minimum to its maximum value. The throttling range is usually adjustable from 2 to 10°F (1 to 5°C).

Thermostats do not have setpoints in the sense of having a predetermined temperature at which they would set the controlled space. Integral action is needed for a controller to be able to return the measured variable to a setpoint after a load change. T_O does not represent a setpoint. It identifies the space temperature at the mid-point of the operating line. This is called the normal condition because at this point the thermostat can both increase and decrease the control output as space temperature changes.

If this controls a cooling damper and the cooling load doubles, the damper will need to be full open, which cannot take place until the controlled space temperature goes to the upper end of the throttling limit.

When the throttling range is narrow enough, the controller is better at keeping the variable near a setpoint. But, the narrow range only allows the vari-able to drift within smaller limits.

There are special thermostats that operate at different normal temperature values for day and night. They have both a day and a night setting dial. The pneumatic day-night thermostat uses a two-pressure air supply system. Changing the pressure at a central point from one value to the other actuates switching devices in the thermostat and indexes them from day to night or night to day. In electric units dedicated clocks and switches are built into each thermostat.

Heating/cooling or summer/winter thermostats can have their action reversed and can have their setpoints changed by indexing. This type of thermostat can be used to actuate controlled devices, such as valves or dampers, that regulate a heating source at one time and a cooling source at another. In a heating-cooling thermostat, there are usually two bimetallic elements, one is direct acting for the heating mode, and the other reverse acting for the cooling mode.

Limited control range thermostats limit the room temperature in the heating season to a maximum of 75°F (24°C), even if the room thermostat is set beyond these limits. This is done internally without using a physical stop on the setting.

A master thermostat measuring outdoor air temperature can be used to adjust a submaster thermostat controlling the water temperature in a heating system. Master-submaster combinations are sometimes designated as single-cascade action. When this type of action is accomplished by a single thermostat with more than one measuring element, it is called compensated control.

Smart Thermostats

Microprocessor-based units with RTD or solid-state sensors are also available. These units usually have their own dedicated memory and intelligence. A communication link is provided to a central computer. The microprocessor-based units can minimize building operating costs by combining features such as time-of-day controls with intelligent comfort gap selection and maximized self-heating.

Zero Energy Band

Available thermostats include the zero energy band (ZEB) types. The ZEB control concept conserves energy by using a comfort gap. A conventional thermostat may continue to use energy when the area's temperature is already comfortable. The comfort gap or ZEB on the thermostat is adjustable and can be changed to match the use of the space.

ZEB control can be accomplished with a single setpoint and single output. (Figure 2-5) The cooling valve will open at the upper end of the range, while the heating valve will open at the lower end of the range. Between these values, both valves are closed and no energy is expended as long as the thermostat output is within this range. The throttling range is usually adjustable from 5 to 25°F (3 to 13°C).

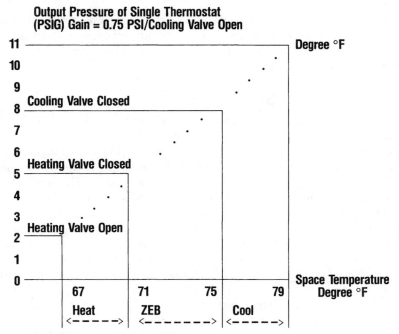

Figure 2-5 *Single setpoint split-range ZEB control*

Although the split-range approach is less expensive than the dual setpoint system (shown in Figure 2-6), it is also less flexible and more restrictive. The gap width can be adjusted only by changing the thermostat gain. The heating valve must fail open, which is a problem for energy conservation.

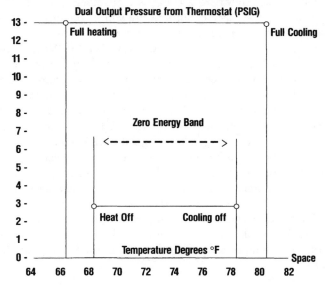

Figure 2-6 *Dual Setpoint ZEB control*

These problems are solved when a dual setpoint, dual output thermostat is used. Both valves can fail closed, and the bandwidth and the thermostat gain are independently adjustable. The gains of the heating and cooling thermostats are also independently adjustable.

Gap Control and Self-Heating

Gap control is an enhancement of zone control. When the comfort level in a zone is between some acceptable limits, the cost of additional energy is not justified. The zones can float between limits instead of trying to hold them at some fixed condition. This approach can substantially reduce the operating cost of the building. The savings come from the use of these acceptable limits which reduces the yearly total of required degree days of heating and cooling.

This is combined with the fact that the building becomes self-heating during winter conditions. During winter conditions, gap control is used to transfer the heat generated in the inside of the building to the perimeter areas, where heating is needed. This can result in long periods of building operation without the use of any fueled energy. The self-heating depends on the winter and summer enthalpy settings in a building. The air returned to the perimeter of the building can add about 10 BTU/lbm which allows the building to be self-heating.

Gap control acts like a type of override mode of control that is superimposed on the operation of the individual zone thermostats. It can be implemented by various techniques, but the flexibility and ease of adjustment of a computerized control makes them ideal in applications where the gap limits of the different zones are likely to change frequently.

Consider the comfort zones defined by combinations of temperature and humidity conditions. Anywhere within these envelopes is deemed to be comfortable. As long as the space conditions are within the envelope, there is no need to spend money or energy to change those conditions.

This comfort gap is a ZEB with humidity control. When a zone of the airhandler is within this comfort gap, its reheat coil is turned off and its variable air volume (VAV) unit is closed to the minimum flow required for air refreshment. When all the zones are inside the ZEB, hot water, chilled water and steam supplies to the airhandler are closed and the fan is operated at minimum flow. When other airhandlers are also within the ZEB, pumping stations, chillers, cooling towers, and hot water generators are also turned off.

In larger buildings that have interior spaces that are heat-generating in the winter, ZEB control can make the building self-heating. Optimized control systems in use today are transferring the interior heat to the perimeter without fueled heat until the outside temperature drops below 10 to 20°F (−12.3° to -6.8°C).

In regions where the winter temperature does not drop below 10°F (−12.3°C), ZEB control can eliminate the need for fueled heat completely. In regions with lower temperatures, ZEB control can lower the yearly heating fuel bill by 30 to 50%.

Supply Air Temperature

One source of inefficiency in many HVAC control systems is the improper arrangement of temperature controllers. Several separate temperature control loops in series may exist.

If one of these uncoordinated controllers is used to control the mixed air temperature, another to maintain the supply air temperature, and another to control the zone-reheat coil, there may be simultaneous heating and cooling periods.

This waste can be eliminated using a coordinated split-range temperature control system, such as that outlined in Figure 2-7. This system can reduce yearly operating costs by more than 10%.

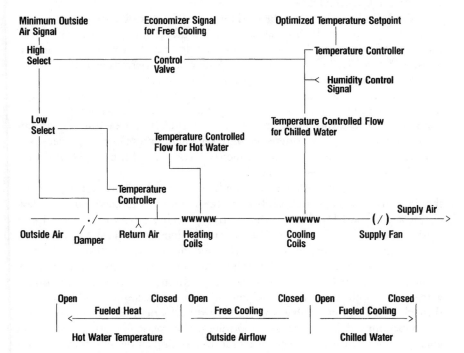

Figure 2-7 Coordinated split-range temperature control

In this type of system, the supply air temperature setpoint is set by a temperature controller which is continuously modulated to follow the load. The loop automatically controls all of the heating or cooling modes.

When the temperature controller output signal is low, heating takes place. As the output signal increases, heating is terminated. If free cooling is available, it is initiated by controlling the outside air damper. When this damper is fully open, the cooling potential represented by free cooling is exhausted and pay cooling is started by opening the chilled water valve. In such split-range systems, the possibility of simultaneous heating and cooling is eliminated. Also eliminated are interactions and cycling.

An override is used to limit the allowable opening of the outside air damper so that the mixed-air temperature will never drop to the freezing point and cause freeze-up of the water coils. The minimum outdoor air requirement signal guarantees that the outside airflow will not be allowed to drop below this limit.

The economizer signal opens the outside air damper when free cooling is available. The possibility for free cooling exists when the enthalpy of the outdoor air is below that of the return air. The humidity control will override the temperature control signal when the need for dehumidification requires that the supply-air temperature be lowered below the setpoint.

Humidity Controls

Wet-bulb thermostats can be used for humidity control since the difference between wet- and dry-bulb temperature will indicate the moisture content. A wick or other means for keeping the bulb wet and rapid air motion are needed to assure a true wet-bulb measurement. A dew point thermostat is designed to control humidity based on the dew point temperatures.

Humidity in the zones is controlled according to the moisture content of the combined return air. The process of humidity control is generally slow and contains large dead-time and transport-lag elements. A change in the supply air humidity will usually not be detected until several minutes later. To prevent possible saturation of the supply air, the moisture contents of the supply air should never be allowed to exceed 90% RH.

For best operating efficiency, a nonlinear controller with a neutral band should be used. This neutral band can be set to a normal range of humidity levels (30% to 50% RH). Then if the relative humidity stays within these limits neither humidification nor dehumidification is demanded. This can lower the cost of humidity control during the spring and fall seasons by 20%.

The same controller can control both humidification, through the relative humidity control valve, and dehumidification, through the temperature control valve, on a split-range basis. Dehumidification is accomplished by cooling through a chilled water valve which is controlled by humidity or temperature.

Subcooling the air to remove moisture can increase operating costs if this energy is not recovered. The penalties for overcooling for dehumidification purposes include the high chilled water cost and the possible need for reheating at the zone level. A pump around economizer can eliminate 80% of this waste. (Figure 2-8)

When the chilled water valve is open more than is necessary, the pump around economizer is started. One economizer coil reheats the dehumidified supply air, using the heat that the pump-around loop removed from the outside air in the other economizer coil before it entered the main cooling coil.

This reduces the chilled-water demand in the main cooling coil and eliminates the need for zone reheating. Besides using a modulated controller setting the speed of the circulating pump, it is also possible to use a constant-speed pump operated by a gap switch.

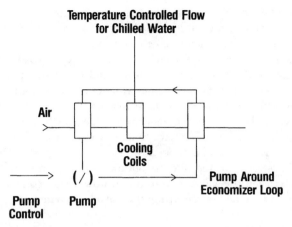

Figure 2-8 Pump around economizer

Outdoor Air Control

Outdoor air is used to satisfy the requirements for fresh air and to provide free cooling. This control is shown in Figure 2-9. Enthalpy logic is used to compare the enthalpies of the outside of return air. An economizer cycle is used when the outside air enthalpy is less than the return air enthalpy.

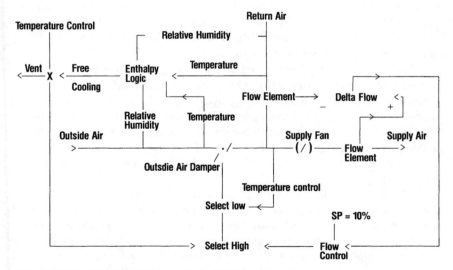

Figure 2-9 Outside air control loops for fresh air or free cooling

The minimum requirement for fresh outdoor air while the building is occupied is usually about 10% of the airhandler's capacity. In an advanced control system, this value will not be a fixed percentage but will vary as a function of the number of people in the building or carbon dioxide in the air.

In most conventional systems the minimum outdoor air is provided by keeping 10% of the area of the outdoor air damper always open when the building is occupied. This method is faulty since a constant damper opening does not provide a constant airflow. The flow will vary with the fan load, since changes in load will change the fan's suction pressure and affect the differential pressure across the damper. This results in some waste of air conditioning energy at high loads and insufficient air refreshment at low loads.

An advanced control system reduces the operating costs while maintaining a constant minimum rate of air refreshment, which is unaffected by fan loading. The direct measurement of outdoor airflow may not be possible because of space or other limitations. In this case the outdoor airflow can be determined as the difference between the supply and return airflows. The required minimum outdoor airflow is obtained by throttling the outside air damper.

CO_2 Ventilation

In older installations the amount of outdoor air used is often based on one of the following guidelines:

1. Outdoor air = 10 to 25% of total air supply rates.
2. Outdoor air = 0.1 to 0.25 cfm/feet (30 to 76 lpm/m) of floor area.
3. Outdoor air = Approximately 5 cfm (25 lps) volumetric rate per person.

These guidelines were formulated when energy conservation was not an important consideration. Their goal was to provide simple, easily enforceable rules that would guarantee that the outdoor air intake exceeds the required minimum. Today the goal is almost the opposite; air quality at minimum cost. Since unoccupied rooms do not need oxygen, some of these rules do not apply.

The relationship between savings in building operating costs and the reduction in outdoor air admitted into the building is strong. In the United States the infiltration of outdoor air accounts for over half of the total heating load and approaches half of the total cooling load. One study showed that 75% of the fuel oil used in New York City schools was consumed by heating ventilated air. Since building conditioning accounts for about 20% of the energy consumed in the United States, the optimized use of outdoor air can be a major contribution in saving energy.

CO_2-based ventilation controls can be a major factor in this savings. The purpose of ventilation is to maintain a certain air quality in the conditioned space. Smoke, odors, and other air contaminants can be correlated to the CO_2 content of the return air. This represents a powerful tool for optimization. The amount of outdoor air required for ventilation can be determined from CO_2 measurements and the time of admitting this air can be selected so that the air addition will also be energy efficient. This technique frees health and energy considerations from conflict and they actually complement each other. CO_2-based ventilation controls can be integrated into the economizer cycle and may be implemented by use of conventional or computerized control systems.

Since the rate of CO_2 generation by a sedentary adult is 0.75 cfh (27 lph), building control by CO_2 concentration will automatically reflect the level of building occupancy. Energy savings of 40% are possible by converting conventional ventilation systems to intermittent CO_2-based operation.

Economizer Cycles

The maximum use of free cooling can reduce the yearly air conditioning load by more than 10%. Enthalpy comparison is used to operate the outdoor air damper when free cooling is available. This type of economizer cycle is activated when the enthalpy of the outdoor air is below that of the return air.

Free cooling can also be used to advantage when the building is unoccupied. Purging or flushing the building with cool outdoor air in the early morning hours results in some cooling capacity being stored in the building structure which reduces daytime cooling loads.

Conventional economizers determine the enthalpy of the outdoor and return air streams, using marginally accurate sensors. They do consider the free cooling potential of the outside air, but there are other possibilities of using the outdoor air to optimize the system.

Advanced, microprocessor-based economizer control can solve these problems. The use of more accurate sensors and the storing in memory the psychrometric chart can be used to maximize all of the potential uses of outside air not just free cooling. Zones of operation based on the relative state of the outside and return air are used.

When the enthalpy of the outside air falls in a zone, its BTU content is too high and no free cooling is available. This is the summer region where the use of outside air should be minimized.

In the winter or fall the enthalpy of the outside air on sunny afternoons may exceed the return air enthalpy. Under these conditions, free heating can be obtained by admitting the outside air.

If the outside air of a zone has a low enough BTU content with a temperature of less than 78°F free cooling is available. If the outside air temperature of a zone exceeds 78°F free dehumidifying (latent cooling) may be available.

When there is a dual duct system in the building with both cold and hot air ducts the zone control system will operate differently depending on the outside air temperature. If it is in the temperature range of the zone, a maximum of 100% outside air can be used. If it is below that temperature, the use of outside air may need to be modulated or time-proportioned.

If the outside air is cooler than the cold-deck temperature and free cooling may be available, but the outside air damper must be cycled to admit enough cold outside air. This air should be sufficient to lower the mixed air temperature to equal the cold-deck temperature setpoint. This will eliminate the need for fueled cooling.

If the outside air is hot and dry, the control system should admit some of it to reduce the overall need for latent cooling. The amount of outside air admitted under these conditions should be controlled so that the resulting mixed-air dew point temperatures will equal the desired cold-deck dew point. This takes advantage of the free drying potential of the outside air. The use of these economizer cycles is akin to opening the windows in a room to achieve a better comfort level.

Optimizing Strategies

The goal of optimization in airhandlers is to meet the demand for conditioned air on a continuous, flexible, and fuel efficient basis. This can require floating of air pressures and temperatures to minimize fuel costs while closely following a changing load.

In many HVAC control systems, the setpoint is set manually and is kept constant. This is wasteful since for each load distribution the optimum supply air temperature is different. When the manual setting is less than the temperature, some zones may become uncontrollable. If the setting is over the optimum, energy is wasted.

Winter Optimization

In winter the main goal of optimization is to distribute the heat load between the main heater and the zone heaters in the most efficient way. The highest efficiency is obtained when the supply air temperature is high enough to meet the load of the zone with the minimum load. This allows the zone reheat coils to only provide the difference between the loads of the various zones.

The setpoint of the least open valve loop is adjusted to keep the valve about 10% open. If the valve opening is more than 10%, the setpoint is increased. If it is less, the setpoint is lowered. Stable operation is possible by making the loop an integral-only control with an external feedback to prevent reset wind-up.

The hot water supply temperature should be independently adjustable and modulated by the control valve so that it always returns to 90% open. Optimizing this temperature minimizes heat-pump operating costs since the temperature is minimized. This also reduces pumping costs by opening all temperature control valves in the system which eliminates unstable cycling in valve operations. Cycling can occur when the valves are in the nearly closed position.

If the hot water supply temperature cannot be modulated to keep the most-open valve from opening to more than 90%, then a slightly different control loop should be used. In this loop, as long as the most-open valve is less than 90% open, the supply air temperature is set to keep the least-open valve at 10% open.

When the most-open valve reaches the 90% opening point, control of the least-open valve is dropped and the loop control keeps the most-open valve from becoming fully open. This is called herding control, where a single constraint envelope herds the valve openings to be within an acceptable band in order to follow the load.

Summer Optimization

In the cooling mode used during the summer, the supply air temperature is modulated to keep the most-open variable volume box from fully opening. If a control element is fully open, it can no longer control. The occurrence of the fully open state must generally be avoided. Sometimes, it is desirable to open throttling devices such as VAV boxes. You can eliminate cycling and unstable operations which are more likely to occur when the VAV box is nearly closed. You can also reduce the friction drop in the system and allow the airhandler to meet the load at the highest possible supply-air temperature.

If the system has undersized ducts and/or fans, the goal of optimization becomes the movement of the minimum quantity of air. The amount of air required to meet a cooling load can be minimized if the cooling capacity of each unit of air is maximized. If fan operating cost is to be optimized, the supply air temperature should be kept at its minimum, instead of being controlled.

If fan operating costs are less than cooling costs, a damper position controller with integral action only is used. Its setting should be 10 times the integral time of the temperature controller. External feedback is used to eliminate reset wind-up.

If automatic switching between winter and summer modes is required, a control system that optimizes the water and air temperatures in both summer and winter is needed. This control system will operate with maximum energy efficiency during the transitional periods of fall and spring. The high efficiency results from the self-heating effect.

Some zones may require heating (perimeter offices) while others require cooling (interior spaces), the airhandler can transfer the free heat from the interior to the perimeter zones by intermixing the return air from the various zones.

A 10 degree ZEB can be used. When the zone temperatures are within the comfort gap of 68°F (20°C) to 70°F (26°C), no fueled energy is used and the airhandler is in a self-heating, or free-heating mode. This results in reducing operating costs and the savings can be more than 30% during the transitional seasons.

Auto-Balancing

A computerized control systems allows greater optimization potentials than non-computerized systems. When the zone conditions are detected and controlled by computer, it can optimize the start-up as well as the normal operation of the building. The optimization of airhandler fans involves two goals. First, the optimum value for the setpoint of the supply air pressure controller is computed. Generally, the supply air pressure is optimum when it is at the lowest possible value, while all loads are satisfied.

When the supply pressure is lowered, fan operating cost is reduced. The lower supply pressures force the VAV boxes for the individual zones to open up so the airflow to the zones is not reduced. The optimum setting for the pressure controller is the pressure where the most-open VAV box is almost 100% open, while the other VAV boxes are less than 100% open.

The other goal of optimization involves rebalancing the air distribution in the building as the load changes. When the VAV boxes are not pressure independent and are not able to maintain constant airflow when the supply pressure changes, manual rebalancing is needed every time the load distribution changes. This is time-consuming and inefficient.

One type of optimization strategy for automatic rebalancing depends on finding the optimum setpoints for the supply air pressure and temperature. The upper limit of the comfort zone is defined as the maximum allowable zone temperature. The lower limit of the comfort zone is defined as the minimum allowable zone temperature. The zone temperatures are checked five minutes after start-up and ten minutes after start-up.

Start-Up Program Flow

1. All VAV boxes are set to the minimum openings required for ventilation (VMIN), such as 25%.
2. The supply air pressure controller is set to midscale so its setpoint is 50%.
3. The temperature controller is set for the minimum allowable zone temperature +25°F in the heating mode and for the maximum allowable zone temperature −25°F in the cooling mode.
4. Every 5 minutes the zone temperatures are checked and every 10 minutes the supply air and return air temperatures are checked and used to build a logic base as shown in Table 2-2.

Table 2-2
Startup Openings for VAV Boxes

	Supply and Zone Temperatures	Temperature Change	VAV Opening %
Heating			
AT > RT	TL− ZT10 > 5°F	ZT10 − ZT5	
Cooling			
AT < RT	ZT10 − TH > 5°F	ZT5 − ZT10	
		< 0.5°F	100
		0.5 − 1°F	90
		1 − 1.5°F	80
		1.5 − 2°F	70
		2 − 3°F	60
		3 − 4°F	50
		4 − 5°F	40
		> 5°F	30
		< 5°F	25

This logic table determines the startup openings of the VAV boxes. When the zone temperature after 10-minutes of operation is within 5°F of the comfort zone, the VAV box can be left at its minimum opening. If the temperature is not within 5°F, a higher opening is required.

The initial VAV opening is changed based on the zone performance during the previous 5-minutes. A larger temperature change in the zone for the previous 5-minutes means it will reach the comfort zone sooner and a smaller opening is required. The VAV boxes on the zones that are farthest from comfort and moving most slowly toward comfort will have the highest openings.

VAV Throttling

The initial VAV opening for each zone is used as the maximum limit on the VAV opening for the first 5-minutes of normal operation. This limit is revaluated every 5-minutes and adjusted based on the recent history of the VAV openings using Table 2-3. If a VAV damper has been throttled in between the MAX and MIN limits, no changes will be made.

Table 2-3 Maximum VAV Opening Flow

VAV condition		Output
Continuously open to its maximum for the last 5-minutes?	Continuously throttled to its minimum for the last 5-minutes?	Change in maximum valve opening at the end of 5-minute period?
Yes	Yes	Leave XMAX = XMIN ?
	No	Increase by 10%
No	Yes	Decrease by 10%
	No	Leave as is

This scheme keeps changes in the load distribution from starving some zones and the building is automatically rebalanced. The minimum openings are determined by the ventilation requirements. The individual VAV openings are checked every few minutes as shown in Table 2-4. Optimizing and auto-balancing is based on the algorithm that whenever a zone is inside the comfort gap, its VAV opening is reduced to a minimum value. This reduces the load on the fans and also provides more air to the zones with the highest loads.

Table 2-4 Openings of VAV Boxes

Operating Mode	Control Criteria	Required VAV openings:
Heating AT > ZT	$ZT < TL - 1$	Maximum
	$TL - 1 < ZT < TL+1$	No change
	$ZT > TL + 1$	Minimum
Cooling AT < ZT	$ZT > TH + 1$	Maximum
	$TH - 1 < ZT < TH + 1$	No change
	$ZT < TH - 1$	Minimum

Air Supply Pressure and Temperature Optimization

The cost of operating the airhandler depends on the cost of air movement (transportation) and conditioning. These two cost factors affect each other. Minimizing the cost of one will increase the cost of the other.

Both the transportation and the conditioning costs should be monitored continuously. Then the larger value can be minimized. A computerized control system allows these costs to be calculated based on the current utility costs and quantities.

If the transportation cost exceeds the conditioning cost, the goal should be to minimize fan operation. This is done by conditioning the space using as little air as possible.

The quantity of air transported is minimized by forcing each pound of air to transport more conditioning energy. Each pound of air will carry more cooling or heating BTUs. The air supply pressure should be as low as possible. The supply temperature is maximized in the winter and minimized in the summer. Fan costs usually exceed the conditioning costs when the loads are low, in the spring or fall, or when an economizer cycle is used to provide free cooling.

Table 2-5a summarizes this algorithm. If none of the VAV boxes are fully open, than all loads are satisfied. The air pressure controller setpoint is kept at a minimum, and the air temperature controller setpoint is lowered in the winter and raised in the summer. If more than one VAV box is fully open, the air supply temperature is increased in the winter and lowered in the summer.

Table 2-5a Optimization of Supply Air Pressure and Temperature,
Spring and Fall, Fan Costs > Conditioning Costs
Sample time = 5 minutes

VAV Boxes	Airhandler Mode	TIC SP @ its limit?	Incremental Ramp Adjustment for Setpoints	
			TIC	PIC
	Heating AT > RT	Yes, @ max.	-2°F	NC, @ minimum
		No	-1°F	NC, @ minimum
None at 100% for 15-minutes				
	Cooling AT < RT	Yes, @ min.	+2°F	NC @ minimum
		No	+1°F	NC @ minimum
< than 1 @ 100% for 30-minutes	Heating AT > RT	Yes, @ max.	NC	NC
		No	NC	NC
	Cooling AT < RT	Yes, @ min.	NC at max.	NC
		No	NC	NC
>1 @ 100% for > than 30-minutes	Heating AT > RT	Yes, @ max.	@+ 25 inch H_2O max.	
		No	+ 1°F	NC
	Cooling AT < RT	Yes, @ min	@ min.	+ 25 inch H_2O
		No	-1°F	NC
> one @ 100% for > 60 minutes	Heating AT > RT	Yes, @ max.	@ max.	+0.5 inch H_2O
		No	+2°F	NC
	Cooling AT < RT	Yes, @ min.	@ min.	+0.5 inch H_2O
		No	-2°F	NC

*NC = No change is made at the end of the 5- minute sampling period

Another algorithm is used when the conditioning costs are higher than the fan operating costs.(Table 2-5b)

Table 2-5b Optimization of Supply Air Pressure and Temperature,
Summer/Winter, Conditioning Costs > Fan Costs
Sample time = 5 minutes

| | | | Incremental Ramp Adjustment for Setpoints | |
VAV Status	Airhandler Mode	PIC Setpoint @ maximum?	TIC	PIC
	Heating AT > RT	Yes	-1°F	-.5 inch H_2O
None @ 100% for 15 minutes		No	NC, @ minimum	-.25 inch H_2O
	Cooling AT > RT	Yes	+1°F	-.5 inch H_2O
		No	NC, at maximum	-.25 inch H_2O
Not > one @ 100% for > 30-minutes	Heating AT > RT	Yes	NC	NC, @ maximum
		No	NC	NC
	Cooling AT < RT	Yes	NC	NC, @ maximum
		No	NC	NC
> one @ 100% for > 30-min.	Heating AT > RT	Yes	+1°F	@ maximum
		No	NC	+.25 inch H_2O
	Cooling AT < RT	Yes	-1°F	@ maximum
		No	NC	+.25 inch H_2O
> one @ 100% for > 60-minutes	Heating AT > RT	Yes	+2°F	@ max
		No	NC	+.5 inch H_2O
	Cooling AT < RT	Yes	-2°F	@ maximum
		No	NC	+.5 inch H_2O
NC = No Change				

This is usually true when the load is high, such as in the summer or the winter. In this case the supply pressure is maximized before the supply air temperature is increased in the winter or lowered in the summer. If none of the VAV boxes are fully open, the pressure controller setpoint is lowered and the temperature controller setpoint is set at or near minimum, in the winter and set at maximum in summer. If more than one VAV box is fully open, the pressure controller setpoint is increased to its maximum setting. When the maximum setting is reached, the supply temperature will be increased in the winter or decreased in the summer.

With automatic balancing and minimum operating cost, there is no need for manual labor and operating costs can be reduced by 30%. There is also the flexibility of assigning different comfort envelopes to each zone. As the occupancy or building use changes, the comfort zones can be changed automatically.

Chimney Effects

High-rise buildings create a natural draft due to chimney effects which pulls in ambient air near the ground elevation and discharges it at the top of the building. Eliminating the chimney effect can lower the operating cost by about 10%. This requires pressure control. A reference riser is used, which allows all pressure controllers in the building to be referenced to the barometric pressure of the outside atmosphere at a select elevation. This pressure reference allows all zones to be operated at 0.1 inch H_2O (25 Pa) pressure. Constant pressure is maintained at both ends of all elevator shafts.

If the space pressure is the same on the different floors of the building, there will be no pressure gradient to initiate the vertical movement of the air and chimney effects are eliminated. This technique also eliminates drafts or air movement between zones which also reduces the dust contents of the air.

If the building is maintained at a constant reference pressure at ground elevation plus 0.1 inch H_2O, there are higher pressure differentials on the higher floors, when the barometric pressure on the outside drops. Air losses from out-leakage and pressures on the windows will both rise. If the pressure reference is taken at some elevation above ground level, these effects will be reduced on the upper floors, but on the lower floors the windows will be under positive pressure from the outside and air infiltration will be experienced.

Cascaded fan controls can be used where the setpoints of the cascade pressure control slaves are programmed so that the air pressure at the fan is adjusted as the square of flow. The pressure at the end of the distribution headers remains constant. This approach provides the most efficient operation of variable-air volume fans.

Chapter 3
Boiler and Pump Control

Boiler Efficiency

The efficiency of a steam generator is defined as the ratio of the heat transferred to the water (steam) to the heating value of the fuel. The purpose of optimization is to maximize the boiler efficiency as variations occur in the load, fuel, ambient, and boiler conditions. Boiler efficiency is influenced by several factors. Larger boilers that have been cleaned and tuned can be expected to have the following efficiencies: 88% on coal (4% excess oxygen), 87% on oil (3% excess oxygen) and 82% on gas (1.5% excess oxygen). Boiler efficiencies seldom exceed 90% or drop below 65%. These numbers do not include blowdown losses or pump and fan operating costs.

Efficiencies vary with individual design, loading, excess air, flue gas temperature, and boiler maintenance. Fuel savings in smaller boilers can be achieved using the techniques shown in Table 3-1.

Neglecting many of the lesser factors, the efficiency can be calculated using the following equation:

$$E = \frac{100\ [1 - 10^{-3}\ (0.22 + K''y)\ (Ts - Ta) - Hc'']}{1 - y/0.21}$$

where

y	=	mole fraction of oxygen in the flue gas
K''	=	fuel coefficient, 1.01 coal, 1.03 oil, 1.07 natural gas
Ts	=	stack temperature
Ta	=	ambient temperature
Hc''	=	fuel coefficient, 0.02 coal, 0.05 oil, 0.09 gas

Boiler Instrumentation

Typically, steam boilers use a drum-type construction. Only very large, supercritical pressure boilers are the once-through type and are found only in the largest electric generating plants. The boiler may be controlled by pressure or flow, depending upon requirements. The load on a steam boiler refers to the amount of steam demanded by the users.

The boiler must be able to follow the demand for steam and the control system must be capable of satisfying changes in the load. Load changes can be the result of changing requirements or cycling.

The utility must have sufficient turn-down to stay in operation at reduced capacities as portions of the plant are shut down. This consideration usually leads to a greater turndown requirement for the boilers than for any other portion of the plant. In addition to this, it is desired to have maximized boiler efficiency.

Table 3-1 Saving Fuel in Boilers

Item	Possible Cost Savings[1]	Comments
Install thermostatic radiator.	10-15%	One pipe-system easier.
Weather-responsive control for outdoor reset.	5-25%	Hot water boilers only Savings highest with setback thermostat and thermostat radiator valves.
Replace fuel nozzle.	2-10%	Easier with gas. Does not apply to steam systems.
Flame-retention head oil burner[2].	10-25%	Oil-fired boilers only. Better but more costly than downsizing nozzle.
Replace pilot light with electric ignition.	5-10%	Gas boilers only.
Install automatic vent damper.	3-15%	Closes off flue when boiler stops firing to reduce standby heat loss up the flue.
Install a clock thermostat.	10-20%	Automatically lowers temperature.
Install gas power burner.	10-20%	Converts oil and coal systems to gas.

1 Savings are not cumulative.

2 May require downsizing the combustion chamber.

Sensors

The normal control requirements for steam boilers are as follows:

- Steam pressure within +/– 1% of desired pressure,
- Fuel-air ratio within +/– 2% of excess air, +0.4% of excess oxygen, based on a desired "load" versus "excess air" curve,
- Steam water level within +/– 1 inch of desired level,
- Steam temperature within +/– 10°F (5.6°C) of desired temperature and
- Boiler efficiency within +/– 1%.

In order to reach these goals, accurate sensors are needed and the load should not change more than 20% of full scale per minute. The boiler design cannot limit this ability. The most important sensors are the flow detectors which provide the basis for both material and heat balance controls. For the control of the air-fuel ratio, combustion airflow measurement is important.

In older units it was difficult or impossible to obtain ideal flow detection conditions. The usual practice was to provide some device in the flow path of combustion air or combustion gases and field-calibrate it by running combustion tests on the boiler.

These tests were run at various boiler loads. They used fuel flow measurements which were direct or inferred from the steam flow. Measurements of percent of excess air by gas analysis were also used along with combustion equations to determine the airflow.

The airflow measurement is normally calibrated on a relative basis. Flow versus differential pressure characteristics, compensations for variations in temperature, and variations in the desired excess air as a function of load are included in the calibration. The desired result is to have the airflow signal match the steam or fuel flow signals when the desired combustion conditions are reached. The following pressure differential sources are normally considered:

- Burner differential, windbox pressure minus furnace pressure;
- Boiler differential, differential across baffle in combustion gas stream;
- Air heater differential, gas side differential;
- Air heater differential, air side differential;
- Venturi section or flow tube installed in stack;
- Piezometer ring at forced draft fan inlet;
- Venturi section of forced draft duct;
- Orifice section of forced draft duct and
- Air foil section of forced draft duct.

The last four are the most desirable, since they use a primary element designed for flow detection and measure the flow on the clean-air side. None of these sensors meets the dual requirement of high accuracy and rangeability. They are of little value at 30% flow or less. Table 3-2 lists some improved flow sensors. Devices such as the multipoint thermal flow probe or the area-averaging pitot stations represent major advances in combustion airflow detection.

Table 3-2 Boiler Flow Sensor Errors

Flow Stream	Flowmeter	Error (% of flow)			Rangeability
		10%	33%	100%	
Fuel oil	Coriolis mass	0.5	0.5	0.5	20:1
Steam and water	Vortex shedding	1-1.5	1-1.5	1-1.5 (steam)	10:1
		0.5	0.5-1	0.5-1 (H_2O)	
	Orifice	not usable	2-5	0.5	3:1
Air	Air averaging pitot traverse station	not usable	2-10	0.5-2	3:1
	Multipoint thermal	5-20	2-5	1-2	10:1
	Piezometer ring, orifice, Venturi, airfoil	not usable	3-20	2-3	3:1

As the flow is reduced, the error increases in all types except the first two. In the orifice or pitot sensors, the error increases and causes these devices to become useless. This can be reduced by the use of two d/p cells on the same element or with the use of smart units.

If the boiler efficiency is to be monitored based on time-averaged fuel and steam flows, the lowest error that can be achieved is about 1%. The air-fuel ratio cannot be measured to a greater accuracy than the airflow and at high turndown ratios, this error can be high. Both the accurate measurement and the precise control of airflows is essential in boiler optimization.

Boiler Operation Monitoring

Steam pressure indication may be obtained using pressure transducers to generate an electrical resistance signal. The heated water temperature in hot water boilers can be measured by a temperature measuring device such as a RTD.

Boiler burner capacity indication may be obtained from a damper position indicator, connected to the linkage controlling the fuel valves and combustion air dampers. The damper position indicator will provide an electrical resistance signal which is calibrated to give the percentage of full fuel input to the burner.

Drum Level Controls

The steam drum is a major part of the boiler (Figure 3-1). It's main function is to provide a surface area and volume near the top of the boiler where separation of steam from water can occur. It also provides a location for chemical water treatment, addition of feedwater, recirculation water, and blowdown, which removes residue and maintains a specified impurity level to reduce scale formation. These functions all involve the addition and loss of material, resulting in continual volume changes in the drum. Controlling these functions while maintaining the correct water-steam interface level is critical in maximizing boiler efficiency.

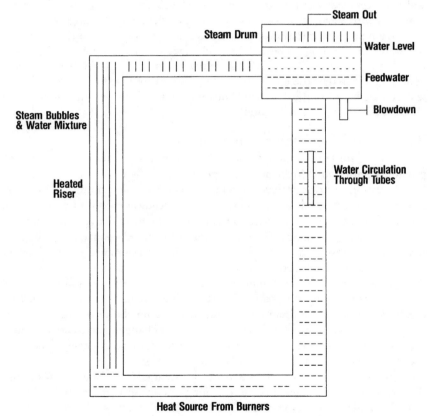

Figure 3-1 Boiler and Steam Drum

Low water level affects the recirculation of water to the boiler tubes and reduces the effectiveness of water treatment. High level reduces the surface area and can lead to water and dissolved solids entering the steam distribution system.

This level is subject to disturbances due to steam pressure fluctuation. As the steam pressure changes with demand, there is a change in level due to the pressure on entrained steam bubbles below the steam drum level. As the pressure drops, there is a rise in level, called swell, that occurs as the trapped bubbles enlarge. As the pressure rises, a drop in level called shrink occurs.

The control requirements are to minimize the drum level variations with steam flow and pressure changes and maximize the surface area in the drum for boiling. Three basic types of drum level control systems are used: single-element, two-element, and three-element.

Single-element Control

The single-element system is the simplest approach. It measures the level and regulates feedwater flow to maintain the level. This system is used for smaller boilers with steady processes and slow and moderate load changes.

Steam pressure variations can cause density changes in both the steam and the water in the drum. These density changes affect the differential pressure (DP) between the variable water head in the steam drum and the fixed reference which is measured by the level transmitter. This means that the actual tank level will not agree with the DP head measurement as the pressure in the tank changes due to the steam demand.

Two- and Three-element Control

If there are larger steam demand fluctuations, a controller is needed that provides continuous correction or compensation of the measured drum level.

A two-element drum level system uses two variables: drum level and the steam flow feedforward signal, to modulate the feedwater control valve. Steam flow load changes are fed forward to the feedwater control valve which provides corrections for load changes. The steam flow feedforward signal can also be compensated for pressure and temperature variations to optimize the feedforward control. This type of control is often used when large variations of steam demand and pressure are common. The steam flow and feedwater flow are matched on a one for one basis. The steam flow signal is combined with the feedback action of the drum level controller to trim the feedwater flow. This tends to compensate for blowdown losses, volume changes, and steam flow measurement errors. This type of control can be used for load changes of moderate speed and magnitude in any size of boiler.

The three-element drum level control system adds feedwater flow control. This requires that the process variable for the control loop be feedwater flow after the boiler is producing steam. This control system allows closer control during transients than the two-element approach. The faster feedwater signal provides an immediate correction for feedwater disturbances.

The drum level is compensated for pressure changes and the steam flow feedforward is used as the process variable only during startup. Once the feed flow is established, it becomes the process variable and the drum level is used as the setpoint.

This type of system can handle large, rapid load changes and feedwater disturbances regardless of boiler capacity. It is often needed on multiple boilers with a common feedwater supply and used for plants where sudden unpredictable steam demand changes can occur. Larger boilers generally use more than one level transmitter for safety and reliability.

Alarms and Safety Interlocks

Those operations relating to safety, startup, shutdown, and burner sequencing are basically digital in nature and operate from inputs such as contact closures. Alarm circuits for emergency conditions include the following:

- Over pressure in steam boiler,
- Over temperature in hot water boiler,
- Fuel gas over/under-pressure,
- Low liquid fuel level and
- Burner flame failure.

A purge interlock will prevent fuel from being admitted to an unfired furnace until the furnace has been completely air-purged. In a low airflow or fan interlock the fuel is shut off upon the loss of airflow and/or the combustion air fan or blower. Fuel is also shut off upon the loss of fuel that could result in unstable flame conditions. All fuel is also shut off when there is a loss of flame in the furnace and fuel to an individual burner is shut off when there is a loss of flame to that burner. A fan interlock stops the forced draft upon the loss of an induced draft fan. Other interlocks that are sometimes used include a fuel shut off on low water level in the boiler or combustible content in the flue gases.

When fans are operated in parallel, an interlock is required to close the shutoff dampers of either fan when it is not in operation. This is needed to prevent air recirculation around the operating fan.

Automatic startup sequencing for lighting the burners and sequencing them in and out of operation is typical. This is done with relay or solid-state, hardwired logic as well as PLCs.

The specification of the various safety interlocks has been influenced by insurance company, NFPA (National Fire Protection Association), and state regulations.

Boiler Time Response

The boiler response to load changes is usually limited by both the equipment design and dead-time considerations. Usually the maximum rate of load change that can be handled is from 20% to 100% per minute. This limitation is due to the maximum rates of change in burner flame propagation and to shrinking and swelling effects on the water level.

The usual period of oscillation of a boiler is between 2 and 5 minutes. This results from a dead time of 30 to 60-seconds. The dead time increases as the load is lowered on a boiler.

Feedforward Loops

One way to reduce dead time and increase the boiler response is to use feedforward loops. A feedforward off steam flow responds to load changes faster than does steam pressure. The induced draft loop can be made more responsive by adding feedforward off the forced draft position. In this system, as soon as the airflow into the furnace changes, the outflow is modified in the same direction, so that the furnace draft pressure is relatively unchanged.

Conventional Air Control

Air controls on traditional boilers have been as inaccurate and unreliable as the airflow sensors when airflow rates were less than 25% of maximum. This is important, since at low loads the boiler tends to be least efficient.

The reasons for this include inaccurate sensors at low flows, leaking, nonlinear dampers with hysteresis and dead bands and the use of constant speed fans.

Many older controls were not based on feedback and used open-loop control. This view neglected nonlinearities, hysteresis, dead time, the play in linkages, sticking dampers, and other effects. In these systems, the nonlinearity of dampers was accounted for characterizing the fuel valve, linkage, and the signal to the fan damper actuator. This was difficult, since louver-type dampers are not only nonlinear, they also suffer from repeatability problems. Multiple-leaf louvers with dividing partitions were used in many of these systems.

The speed of response of the fuel flow control loop is much faster than that of the airflow control loop. Attempts to correct this included making the air actuator twice as fast as the fuel actuator.

Distributed microprocessor-based systems can linearize nonlinear systems. They can also memorize their dead band and speed of response.

Dampers

Dampers are undesirable as final control elements because of their hysteresis, nonlinearity, and leakage. If the flow is accurately measured, these undesirable features are easier to handle. If the damper displays some hysteresis or dead band, a reliable flow sensor allows the opening and closing characteristics of the damper to be determined. If the controls include some digital memory, this characteristic curve can be stored and used later. The damper characteristic curve can also be automatically updated and corrected, as it changes due to wear, dirt buildup, or other causes.

Due to the non-linearity, an increase of 1° in damper rotation that occurs at a 10° opening causes a much greater increase in the actual airflow than a 1° increase that occurs at a 60° opening. This is undesirable since stable controls require constant gain in the loop. Ideally, the change in airflow per incremental change in damper opening should be uniform from the closed to the full open position.

Besides more linear dampers, such as those that use several different sized louvered disks (Mitco type), another solution is to use compensators and characterizer positioners. When the damper gain drops off as it opens, the compensators add more gain into the loop, keeping the total gain nearly constant. These newer damper designs tight shutoff and the efficiency holds up until the load drops to about 10%.

Fans

Damper control wastes fan energy in order to control. A better method of controlling airflow is to eliminate the damper and throttle the fan itself. Instead of using up unnecessary air transportation energy in the form of damper pressure drop, this energy is not introduced. This can be done by the use of variable-speed fans or by varying the pitch of the fan blades.

Variable-speed or variable-pitch fans will conserve air transportation energy and reduce the overall cost of boiler operation. Other advantages include better linearity, hysteresis and dead band. Leakage problems are also reduced.

Pressure Controls

Combustion control can be divided into separate fuel control and combustion air control systems. The relationship between these two systems requires the use of the fuel-air ratio controls. Safety requires that fuel addition should be limited by the amount of available combustion air, and combustion air may need minimum limiting for flame stability.

Feedforward control can improve the pressure control by adjusting fuel as soon as a load change is detected, instead of waiting for the pressure to change.

It may be desirable to have either flow or pressure control, in this case, a master control arrangement, as shown in Figure 3-2 can be used. It may be simpler to switch transmitters, but it is desirable to transfer the controller outputs so that the controller does not have to be retuned each time the measurement is switched.

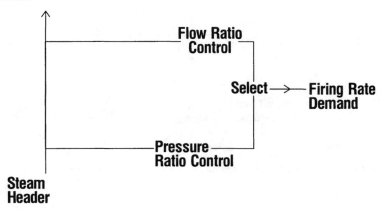

Figure 3-2 Boiler control with alternative pressure or flow control

A typical setting for a pressure-controlled master is 16% for the proportional band and 0.25 repeats per minute for reset. In a flow control master the comparable settings might be 100% for the proportional band and 3.0 repeats per minute for reset.

Feedforward Control of Steam Pressure

In single boiler installations where the steam lines are small, steam line velocities high, and the relative capacity in the steam system low, it may be necessary for control stability reasons to correct the pressure master using feedforward control. This can be done with boiler control using flow-corrected pressure signals. (Figure 3-3) A variation of this uses flow-corrected pressure control with integral correction in the bypass. (Figure 3-4)

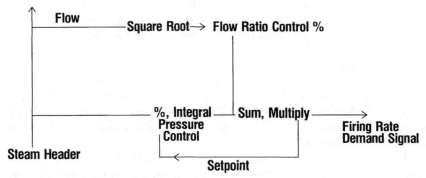

Figure 3-3 Boiler control with flow-corrected pressure

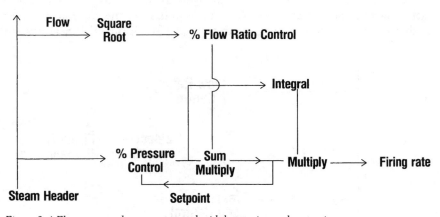

Figure 3-4 Flow-corrected pressure control with bypass integral correction

These control systems allow the use of wider proportional bands and lower reset settings in the pressure controller than would otherwise be possible. The configuration with the integral mode is in the bypass allows it to continuously adjust the fuel and airflow controls to match the steam flow. This provides a quick, stable response to large and rapid load changes.

If either of these control loops is used with more than one boiler, then the steam flow measurement must be the total from all boilers or a highflow selector must be used to select the highest of the steam flows.

Boiler-pressure control can use another type of feedforward control where the fuel flow is set proportionally to the estimate of the load which is equal to the square root of h/p. (Figure 3-5) Here h is the differential developed by a flow element and p is the steam pressure. Dynamic compensation is applied as a lead-lag function to help overcome the heat capacity in the boiler.

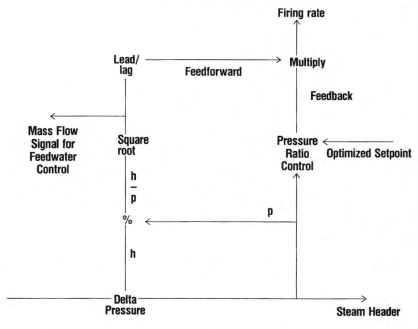

Figure 3-5 Firing rate determination using feedforward loop with feedback trim

A pressure controller adjusts the ratio of the firing rate to estimated load, to correct for inaccuracy of the model and variations in heat combustion of the fuel. A multiplier also changes the gain of the pressure feedback loop proportional to load. This scheme is useful for boiler control at lower loads with their lower velocities.

Two loops measure the pressure and flow of the generated steam. The flow transmitter must be a high rangeability and linear device, capable of accurate measurements at low loads. The firing rate demand signal is generated by the pressure controller.

In order to speed up the response of this loop to load changes, a feedforward trim is used. This trim is based on steam flow, since this flow is the first to respond to load changes.

Fuel Control

The primary boiler fuels are coal, oil, and gas. Gas and oil fuels require the simplest controls, since they are easily measured. A valve positioner which can provide a linear relationship between flow and control signal is desirable but a flow controller is often used to provide more precise linearization. Whenever a 3:1 turndown or better with high measurement accuracy is needed, the Coriolis type of mass flow sensor should be used in place of orifices.

If the fuel is a gas at variable pressure, a pressure control valve is often installed upstream to the flow sensor. Both valves affect the variables, pressure and flow. There will be interactions and to eliminate any oscillations, the following can be done

1. Pressure is unregulated, but pressure-compensate the flow sensor.
2. Assign less pressure drop to the pressure control valve than to the flow control valve.

The latter means using a larger valve for pressure control than for flow control. In order to obtain a more linear relationship between fuel flow and valve position, an equal-percentage valve is needed.

In the use of oil fuels, proper atomization at the burner requires that the oil be kept at constant pressure and viscosity. If heavy oils are burned, they must be circulated past the burner (Figure 3-6). The difference between the outputs of two Coriolis mass flowmeters can indicate the net flow to the burner. The burner backpressure is controlled by the control valve in the recirculating line and the flow controller setpoint is adjusted by the firing rate demand signal. Atomizing steam is ratioed to the firing rate, and the heating steam is modulated to keep the fuel viscosity constant.

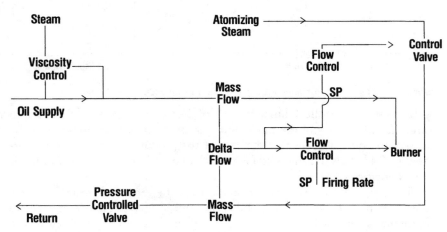

Figure 3-6 Oil flow control

Closed-loop control can provide greater precision and better linearization, but performance can be limited by the hardware or installation. One problem is transmitter rangeability if orifices are used instead of turbine or mass flowmeters. Installation problems include the use of long transmission lines which introduce dead time into the loop if pneumatic leads are installed without boosters.

In any type of fuel control the maximum flexibility is present if all flow signals and control valve characteristics are linear. This allows the various flows and signals to be combined, subtracted, multiplied, or divided to produce the desired control.

Airflow Control

The combustion air for steam boilers can be supplied in the following ways:
- Induced draft, suction fan at boiler outlet or stack draft;
- Forced draft, pressure fan at inlet and
- Combination of above.

The combination is known as balanced draft where a slight negative pressure is maintained in the furnace.

In the control of the air-fuel ratio, combustion airflow measurement is important. Regardless of the type of primary element used, the signal will usually be noisy due to pulsations from the pumping of the fans or the combustion process. Dampening the flow signals is needed before the controller can be tuned.

Damper and Fan Control

The control devices for boiler airflow control are either double-acting pistons or electric motors. A linear relationship is required between the control signal and combustion airflow. Closed-loop air control is more precise and self-linearizing. For a single-damper-controlled open loop a combination of damper and speed control can be used to increase rangeability. When two fans are operated in parallel or on single-fan operation, the idle fan should have its damper closed to prevent recirculation from the operating fan. When one fan of a two-fan system is switched on or off, airflow upsets can occur unless manual or automatically compensation of the operating fan damper position is used. Closed-loop control loops consist of flow controllers with airflow feedback.

Furnace Draft Control

The measurement of furnace draft can result in a noisy signal which limits the loop gain to lower values. In order to provide control without excessive noise, the span should be as large as possible. It may also be necessary to use integral control to stabilize the loop.

Airflow and Furnace Pressure

Stability problems and interactions can occur because of measurement lags. Larger diameter pressure transmitter connections can be used to reduce this lag.

Either the forced or the induced draft fan can be used to control the furnace draft with the other fan performing the basic airflow control function. Interactions cannot be completely eliminated between these two loops, but it can be minimized by the rules of Table 3-3.

Table 3-3 Furnace Draft Control

Forced Draft Fan	Induced Draft Fan
Airflow detected by	Airflow detected by
Boiler differential	Air duct Venturi
Gas side differential pressure	Air duct orifice
of air heater	Air side differential pressure
Venturi in stack	of air heater
	Piezometer ring in forced
	draft inlet
	Area averaging pitot on air side
Airflow control	
Throttle the induced draft fan	Throttle the forced draft fan

Airflow should be measured and controlled on the same side [air or combustion gas] of the furnace to minimize interactions between the flow and pressure loops. If the combustion air is preheated, then its temperature will vary greatly and compensation is needed. The mass flow of air can be related to the square root of h/T, where h is the differential across a restriction and T is absolute temperature. This technique can be used with excess oxygen trim on the airflow controller.

If the dampers are the same size, their pressure drops will be similar and decoupling may be needed. Connecting the two dampers (or fans) in parallel and using the furnace pressure as a trimming signal can be used. This will also overcome problems from noisy furnace pressure signals and slow response caused by the series relationship between flow and pressure loops.

In this arrangement the airflow controls move both dampers equally, and the furnace pressure corrects for any mismatch. The furnace pressure may respond faster to a change in the downstream damper opening and a dynamic lag is required.

The airflow is detected by a high rangeability flowmeter which is compensated for temperature variations to approximate mass flow. The air-fuel ratio is adjusted by applying a gain to the airflow and the optimization of this ratio provides a major part of the boiler optimization. An airflow controller throttles the speed or blade pitch of the forced draft fan and the signal to the induced draft fan using feedforward control.

Dynamic compensation is provided by a lag network or module. As soon as the air inflow to the furnace changes, the outflow should also start changing. This will improve the control of the furnace draft.

Fuel-Air Ratio

The fuel-air ratio should be maintained on a real-time rather than a time-averaged basis. Usually fuel and air should be controlled in parallel rather than in series. This is done because a lag of only one or two seconds in measurement or transmission can seriously upset the combustion conditions in a series system. This can result in alternating periods of excess and deficient combustion air.

Control of the fuel-air ratio can be done with a system calibrated in parallel with provisions for the operator to make manual corrections. The operator must compensate for variations in fuel pressure, temperature, or heating value or for air temperature, humidity, and other factors. Detailed testing at various loads is needed for characterizing and matching fuel and air control devices. These systems also need to be adjusted for higher excess air, since they have no way automatically compensating for the fuel and air variations.

Another control technique is to use proportional compensation. This is done by balancing fuel burner pressure to the differential produced between windbox and furnace pressures.

Burner fuel apertures are used to measure fuel flow, and the burner air throat is used as a primary element to detect airflow. The rangeability is limited unless there are multiple burners that can be put in or out of service. The control can also be reversed, with the firing rate demand adjusting the fuel and the correction control on the airflow.

A variation of this control is to measure flows more accurately and use them in a flow controller to readjust the primary loop through a combining relay. In this system fuel and airflow are open-loop controlled, and only secondary use is made of their measurements. The effects of interaction and disturbances in the fuel and air control loops can be minimized by the use of closed-loop fuel and air control. Field testing is less critical since the self-linearizing closed-loop system reduces the need to characterize fuel and air control devices.

Fuel and Air Limiting

Maintaining the correct fuel-air ratio helps in the following areas:
- Limiting the fuel rate to available air,
- Limiting the air minimum and maximum to fuel flow,
- Limiting air leading fuel on load increases and
- Limiting air lagging fuel on load decreases.

Limiting the fuel to available airflow and limiting the airflow to match the minimum fuel flow provides safety features along with limiting the minimum fuel flow to maintain a stable flame.

These limiting features can be applied with the type of noninteracting, self-linearizing system that uses closed-loop control of the air-fuel ratio. The necessary modifications can provide these features without upsetting the setpoint of the fuel-air ratio.

If the actual airflow decreases below the firing rate demand, then the actual airflow signal is selected to become the fuel demand by a low selector. If fuel flow is at a minimum and the firing rate demands further decreases, the actual fuel flow becomes the airflow demand. The fuel signal is selected if it is greater than firing rate demand signal. The fuel is minimum limited by a direct-acting pressure or flow regulator.

This allows the fuel flow to respond to small increases in firing rate demand without raising the airflow. The airflow can also be decreased slightly if the firing rate demand falls, without having to decrease the fuel flow.

In open-loop systems, fuel and air limiting is more difficult to apply. In one scheme, the fuel setpoint is determined or limited by the actual airflow, and a fuel cutback is necessitated by reduced firing rate demand that takes place at the expense of a temporary fuel-air ratio offset.

Limiting combustion airflow to a minimum or to the rate at which fuel is being burned creates some problems, because while the limit is in force, there must be some provision to block the integral action in the airflow controller.

Another way to limit fuel is to have the firing rate demand directly set the airflow, with the fuel being controlled through a combining relay. In this scheme the final fuel-air ratio is done by the correction of a fuel-air ratio controller. One problem with this system is that with a sudden decrease of firing rate demand, the resulting reduction in fuel flow will occur only after the airflow has been reduced.

Feedwater Control

Feedwater control involves the regulation of water to the boiler. The level of water in the boiler must be maintained within a certain range. A decrease in this level can uncover boiler tubes, allowing them to become overheated. An increase in this level can interfere with the operation of the internal boiler equipment.

At each boiler load there is a different volume of water that is occupied by steam bubbles. As the load is increased, there are more steam bubbles which causes the water to swell or rise. The control of feedwater needs to respond to load changes.

A sudden increase in feedwater flow will collapse some bubbles in the drum and temporarily reduce their formation. So, although the mass of liquid in the system has increased, the apparent liquid level in the drum falls. Equilibrium is restored within a few seconds, and the level will begin to rise.

The initial reaction to a change in feedwater flow tends to be in the wrong direction and is called an inverse response. This causes an effective delay in control action, making control more difficult. Liquid level in a vessel without these thermal characteristics can typically be controlled with a proportional band of 10% or less. This type of control would need a proportional band of about 100% to maintain stability. Integral action may be needed. For small boilers with relatively high storage volumes and slow-changing loads, proportional control may be acceptable. There is some instability as a result of the integration of the swell/load changes that must later be removed.

In larger boilers economic considerations make it desirable to reduce drum sizes and increase velocities in water and steam systems. This makes the boiler less able to act as an integrator to absorb the effects of incorrect or insufficient control.

Sensors such as the conductivity type, which are insensitive to density variations should not be used. Displacement and differential pressure sensors are better since they respond to hydrostatic head.

Feedforward Control

Feedforward can reduce the influence of swell and inverse-response. The feedwater flow should match the steam flow in the absence of action by the level controller. Two differential pressure flowmeters are used with identical ranges and their signals are subtracted. (Figure 3-7) If the two flow rates are the same, the subtractor sends a 50% signal to a flow-difference controller. An increase in steam flow calls for an equal increase in feedwater flow to return the difference signal to 50%.

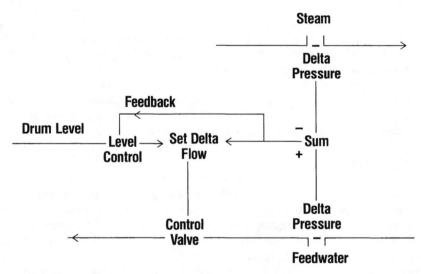

Figure 3-7 Feedforward control of drum level

Errors in the flowmeters and the withdrawal of blowdown water which is not converted to steam will prevent the two flow signals from being identical. Any error in the steam-water balance can cause a falling or rising level. The level controller must readjust the setpoint of the flow-difference controller for a steady-state balance. External feedback from the flow-difference measurement to the level controller preconditions the level controller during startup or when the feedwater is limited.

To keep the controller gain independent of load variations, the feedwater valve should have linear characteristics. One of the best ways to do this is by using a characterizing positioner on the control valve. It may also be necessary to dampen the control valve, to make it less sensitive to noisy control signals. An oversized valve can also degrade the control quality.

Pump Speed Control

Control of pump speed to regulate the feedwater flow may be accomplished when the pump is driven by a steam turbine or variable-speed electric drive. This can be used instead of a control valve to save pump power.

In systems where several boilers are operating in parallel, the control of the discharge pressure at a constant differential pressure above boiler pressure is used. In a single boiler system, most of the speed control range is used at low flow.

Steam Temperature Control

The factors affecting steam temperature include flue gas flow pattern, flue gas mass flow, temperature of flue gases leaving the furnace, and steam flow. Temperature can be controlled by the amount of recirculation of the flue gas.

The required positions of burners or recirculating and bypass dampers as a function of load can be programmed. Adjustments to correct for inaccuracies in the program and changes in the characteristics of the boiler can be done by feedback control of temperature, using proportional and integral action. The feedback loop gain varies directly with load.

Large boilers can have burners that tilt for steam temperature control. This effectively changes the furnace heat transfer area, resulting in temperature changes of flue gases leaving the boiler. To increase the steam temperatures at low loads, recirculation blowers or tilting burners are used.

Loop Interaction

In a complex system, the adding, subtracting, multiplying, dividing, and comparing control signals can make it difficult to get the system on operational control. The systems often interact, airflow affects steam temperature, feedwater flow affects steam pressure, and fuel flow affects drum level and furnace draft. The flows of fuel and air need to be controlled such that the flow rates reaching the burner always represent a safe combination. For flexibility and rangeability, linear flow signals are necessary. Control valves and piston operators need linearizing positioners. Tie-back configurations can simplify the task of getting on automatic control.

Boiler optimization involves the following goals:

- Minimize excess air and flue-gas temperature, o minimize steam pressure,
- Reduce feed pump discharge pressures,
- Reduce heat loss through pipe walls,
- Minimize blowdown and
- Recover condensate heat.

Most of these goals can be achieved by closed-loop control.

Excess Air

When a boiler is operating at a specific load, the boiler losses as a function of air excess or efficiency will have a minimum point. A process with an operating curve minimum is ideal for optimization with improved instrumentation. These control systems continuously determine the minimum point and then shifting the operating conditions to that point.

Most heat losses in a boiler occur through the stack. When there is not enough air unburned fuel leaves, and when there is an air excess, heat is lost as unused oxygen and nitrogen are heated and then discharged into the atmosphere.

The goal of optimization is to minimize the total losses. This is accomplished by minimizing excess air and by minimizing the stack temperature. The fuel-saving potential in a boiler can be in excess of 20%.

The minimum loss point is not where excess oxygen is zero. This is because no burner is capable of providing perfect mixing. This actual minimum loss or maximum efficiency point is found by lowering the excess oxygen as far as possible until opacity or CO readings indicate the minimum has been reached. At this minimum loss point the flue-gas losses balance the unburned fuel losses. The optimum excess oxygen percentages for gas, oil, and coal are about 1%, 2%, and 3%.

Detecting the Flue-Gas Composition

Excess air can be correlated to O_2, CO, CO_2, or the combustibles present in the flue-gas. The combustibles are usually detected either as unburned hydrocarbons or by measuring the opacity. These measurements are not well suited as the basis for optimization, since the goal is not to maintain an optimum concentration, but to eliminate combustibles from the flue gas. These measurements are usually applied as limit overrides.

Excess O_2 as a basis of boiler optimization is a relatively insensitive measurement, but is used with zirconium oxide probes. To minimize duct leakage effects, the probe is installed close to the combustion zone but at a point where the gas temperature is below that of the electrically heated zirconium oxide detector. The flow should be turbulent at the sensor location to ensure that the sample will be well mixed and representative of the flue-gas composition. The output signal of these zirconium oxide probes is logarithmic. The O_2 probe cannot distinguish the oxygen that enters through leakage from excess oxygen left over after combustion.

Another limitation of the zirconium-oxide fuel cell sensor is that it measures net oxygen. If there are combustibles in the flue gas, they will be oxidized on the hot surface of the probe and the instrument will register only that oxygen, which remains after the reaction. This error is not important when the total excess oxygen is about 5%. But, in an optimized boiler, the excess oxygen is only 1% and the difference between total and net O_2 can cause a significant error.

CO Measurement

The best measurement of flue-gas composition is the detection of carbon monoxide. Optimum boiler efficiency is obtained when the losses due to incomplete combustion equal the effects of excess air heat loss. CO should be zero when there is oxygen in the flue-gas. In practice, maximum boiler efficiency can be maintained when the CO is between 100 and 400 ppm.

CO is a sensitive indicator of improperly adjusted burners. If its concentration gets up to 1000 ppm, there may be unsafe conditions. CO is a direct measure of the completeness of combustion and is not affected by air infiltration, except for dilution effects.

Control systems that measure the excess O_2 and carbon monoxide can optimize boiler efficiency, even if load, ambient conditions, or fuel characteristics vary. These systems can detect a shift in the characteristics of the boiler, this can be used to signal the need for maintenance of the burners, heat transfer units, or air and fuel handling equipment.

Nondispersive infrared (IR) analyzers may be used for simultaneous in situ measurement of CO and other gases or vapors such as that of water. This can signal incipient tube leakage.

The CO analyzers cannot operate at high temperatures and are usually located downstream of the last heat exchanger or economizer. At these points, flue-gas dilution due to infiltration is usually high enough to require compensation. The measurement of CO_2 can be used to calculate the compensation needed.

Flue-gas Temperature

Flue-gas temperature is an indicator of boiler efficiency. As it drops below the dew point, a condensate is formed that will dissolve the oxides of sulfur in the flue gas and the acid that results will cause corrosion.

A temperature control loop can keep the flue gas dry and above its dew point. The flue-gas temperature at the cold end of the stack is usually measured at several points in the same plane. When this averaged temperature drops near the dew point, the temperature control loop will start increasing the setpoint of air preheaters. This increases the inlet air temperature to the boiler which raises the flue-gas temperature.

The amount of energy wasted through the stack is a function of both the amount of excess air and the temperature at which the flue gases leave. The flue-gas temperature depends on load, air infiltration, and the condition of the heat transfer surfaces. A stack temperature that rises above baseline indicates a loss of efficiency. Each 50°F (23°C) increase can lower the boiler efficiency by about 1%.

High flue-gas temperatures can cause the following to occur:
• Fouling of heat transfer surfaces in the air preheater,
• Scale buildup on the inside of the boiler tubes,
• Soot buildup on the outside of the boiler tubes and
• Deteriorated baffles that allow hot gases to bypass the boiler tubes.

Microprocessor-based systems can respond to a rise in flue temperature by taking corrective actions. This can include automatically blowing the soot away, or by alarms with detailed maintenance instructions.

When the stack temperature drops below the baseline, this does not always mean an increase in boiler efficiency. It can indicate the loss of heat due to leakage. Cold air or cold water may leak into the stack gases if the economizer or the regenerator are damaged. This can result in a loss of efficiency and problems from corrosive condensation, if the temperature drops to the dew point. A drop in flue-gas temperature also lowers the stack effect and increases the load on the induced draft fan. Both high and low limit constraints on stack temperature should be used.

Fuel Savings and Optimization

The overall boiler efficiency depends on the heat transfer efficiency that is reflected by stack temperature. This should be measured when the excess air is near optimum. Combustion efficiency is related to excess oxygen, which should be as low as possible without exceeding the limits on CO, opacity and unburned carbon and NO_2.

A fuel savings of about 1% can be achieved for each 50°F reduction in stack gas temperature. The total savings from improved thermal and combustion efficiencies due to both stack temperature and excess oxygen being lowered to their limits is about 5%.

Air-Fuel Ratio Control

In older, small boilers the air-fuel ratio was set by mechanical linkages between the fuel control valve, airflow damper and a pressure controller. The setting of the linkages had to be readjusted manually as conditions changed.

Later, the excess oxygen content of the flue-gas was used to provide a feedback trim between the firing rate and damper opening. A high/low limiter was used as a safety precaution to prevent the formation of fuel-rich mixtures from an analyzer or controller failure.

An automatic fuel-air ratio correction can be based on the load and excess air indicated by the percent of oxygen. In this type of control, the relationship between the load (steam flow) and the corresponding excess oxygen setpoint is used for optimum performance. The computer or controller memorizes the characteristic curve of the boiler and generates the excess oxygen setpoint based on that curve.

The closed-loop control is corrected by oxygen analysis and can use safety limits to protect against air deficiency. This system forces airflow to lead fuel on an increasing load and to lag on a decreasing load. The flue-gas oxygen content tends to overshoot the setpoint on all load changes. Integral control is used on the oxygen signal so that reaction to rapid fluctuations is minimized.

A variation of this control system is to use steam flow instead of firing rate as the input and the output is the setpoint of the excess oxygen controller. A bias allows the characterized curve to be shifted to compensate for changes in air infiltration rates or in boiler equipment performance.

The oxygen controller compares the measured flue-gas oxygen concentration to the load-programmed setpoint and applies integral action to correct the offset. Anti-reset windup and adjustable output limiting may also be used. The oxygen controller is direct acting so the air-fuel ratio adjustment factor increases if the oxygen concentration in the stack rises.

Another enhancement involves increasing the speed of response. The transportation lag in a boiler can be as long as a minute. This is the time interval that passes after a change in firing rate before any change can be detected in the composition of the flue gas. This time varies with the flue-gas velocity and usually increases as the load drops. In boilers where loads are frequently changing at irregular intervals, this dead time can cause control problems.

Feedforward control can substantially lower this time. The airflow damper position is controlled with a closed loop. The setpoint is based on a characterization curve developed from the firing rate command. The loop adjusts the excess oxygen using feedforward control.

Feedback trim is provided by an oxygen measurement which changes the setpoint to the damper position controller. This type of system will anticipate the need for excess oxygen changes. It corrects the oxygen concentration to correspond with the excess air curve.

Multivariable Control

Single loop feedback control is primarily a single variable (single input, single output) technique. Each loop is independently operated as a feedback device and no provision is made to compensate for any interaction among the controlled variables.

The standard analog controller has inherent internal interaction. When retuning of a controller becomes necessary, the variable on which the controller is applied may interact with other process variables, which would necessitate retuning these other variables.

The application of a computer to this system with conventional control serves no purpose unless the computer acts to change setpoints and optimizes. When the computer totally replaces the analog conventional control with a digital equivalent, this is sometimes called Direct Digital Control (DDC).

Feedforward/Feedback Control

A disadvantage of feedback control is that the error or deviation must exist before corrective action takes place. Feedforward control is a process where the magnitude of the error is anticipated and corrective action is taken prior to the occurrence of the error. Feedforward/feedback control is a combination of both techniques. The process of combining both concepts results in error anticipation and corrective action followed by a readjustment.

Feedforward components consist of a feed analyzer and feed flow transmitter. These two components analyze and sense the disturbances due to a feed change and composition. If the feed and/or composition changes, the output manipulates the system to approximately the correct or desired quantity. The lead lag functions on the feed flow should be set with respect to the dynamics of the system.

Most industrial control applications over the past half-century have been based only on simple feedback or feedforward/feedback concepts. One method of describing this type of control is single variable or single input/single output control. The requirements of closer constraints, higher throughput and energy saving systems offer a challenge that is difficult to cope with by using only conventional analog control. All of these factors, plus the realization that manipulation of one variable affects others, naturally leads to the study of how these interactions take place and how they can be controlled.

A multivariable control system is a device or set of devices which has some built-in intelligence to simultaneously look at many variable and to choose, based on a given situation, the best of several programmed control strategies. Consider the effects on the control of a boiler of varying the steam flow and other possible variables. Each of these variables commonly affects the others. These effects can be broadly defined as interactive.

Multiple measurements of flue-gas composition can be used to eliminate the need for any manual bias adjustments. They also allow more accurate and faster control than is possible with excess oxygen control alone. The manual bias is replaced by the output signal of a CO controller which trims the characterized setpoint of the excess oxygen controller. This trimming corrects for changes due to incomplete combustion, fuel characteristics, equipment, or ambient conditions.

Other improvements involve correcting the CO measurement for dilution effects due to air infiltration, or characterizing the CO controller setpoint for changing loads. An opacity override can be used to meet environmental regulations. When the setpoint of the opacity controller is reached, it will start biasing the O_2 setpoint upward until opacity returns to normal.

Microprocessor-based systems allow several control variables to be simultaneously monitored. Control is switched depending on which limit is reached. These control systems tend to be both more accurate and faster in response than if control was based on a single variable.

Multivariable envelope-based control can also be implemented by analog controllers that are configured in a selective manner. Each controller measures a different variable and is set to keep that variable at some limit. This is sometimes called herding. Reset windup in idle controllers is prevented with the use of an external reset, which also allows bumpless transfer from one controller to the next.

Steam Pressure Optimization

In traditional boiler operation, the steam pressure was maintained at a constant value. Only recently has this begun to change. In congenerating plants, the boiler steam is used to generate electricity and the turbine exhaust steam is used as a heat source. Optimization is obtained by maximizing the boiler pressure and minimizing the turbine exhaust pressure to maximize the amount of electricity generated.

In plants that do not generate their own electricity, optimization is achieved by minimizing the boiler operating pressure. This lowers the cost of operating feedwater pumps since their discharge pressure is reduced. Pump power in high-pressure boilers is much as 3% of the gross work produced. The control system must find the optimum minimum steam pressure, which then becomes the setpoint. If all steam user valves are less than fully open, a lowering in the steam pressure will not restrict steam availability, since the user valves can open further. A high signal selector chooses the most-open valve, and a valve position controller compares that signal with its setpoint. If even the most-open valve is less than the setpoint value, the pressure controller setpoint is lowered.

The valve position controller is an integral controller. Its reset time should be at least 10 times that of the main or master controller. The slow integral action slows any changes of the steam pressure and noisy valve signals should not occur.

The output signal from the integral controller is limited by a high-low limit control, so that the steam pressure setpoint cannot go outside some preset limits. External feedback is used with the integral controller so that when its output is overridden by a limit there will not be any reset wind up.

This type of optimization, where the steam pressure follows the load, not only increases boiler efficiency, it also has several advantages. It prevents any steam valve from fully opening and thereby losing control. Steam valves are kept away from the unstable (near-closed) area of operation. Valve maintenance is reduced and valve life is increased due to the lower pressure drop.

Efficiency Measurements

Boiler efficiency can be monitored indirectly by the measurement of flue-gas composition, temperature, combustion temperature, and burner firing rate or directly through time-averaged steam and fuel flow monitoring. For the direct efficiency measurement, it is important to use flowmeters with high accuracy and rangeability. The error contribution of the flowmeters should be 1/2 to 3/4%.

The continuous updating and storing of performance data for each boiler is also a valuable tool in operational diagnostics and maintenance.

Pump Optimization

The optimization of a steam generator includes the optimized operation of the feedwater pump at the condensate return system. The two major types of pumps are positive displacement pumps and centrifugal pumps. The centrifugal designs are either the radial-flow or the axial-flow type. Liquid enters the radial-flow designs at the center of the impeller and is forced out by the centrifugal force into a spiral bowl. The axial-flow propeller pumps are designed to push rather than throw the fluid upward. Mixed-flow designs are a combination of the two types.

In a positive displacement pump, a piston or plunger moves inside a cylinder in a reciprocating motion, similar to a one cylinder engine. The stroke length is adjustable within a 10:1 range. This range can be increased to 100:1 with the addition of a variable-speed drive. Plunger types can generate higher discharge pressures than the diaphragm types.

Centrifugal Pumps

Pump efficiency is the ratio of the useful output power of the pump to its input power. Typical pump efficiencies range from 60 to 85%. With a given impeller diameter, pump flow is linearly proportional to pump speed. The pump discharge head is related to the square of pump speed, and the pump power consumption is proportional to the cube of pump speed. These factors allow variable-speed pumps to be very energy efficient.

When liquids are being pumped, the pressure in the suction line should be above the vapor pressure of the fluid. The available head measured at the pump suction is called the net positive suction head (NPSH). The available NPSH represents the difference between the existing absolute suction head and the vapor pressure at the process temperature. If the minimum NPSH is not available, the pump will fail to generate the required suction lift and flow will stop.

Friction losses are related to the square of flow and represent the resistance to flow caused by pipe friction. The friction losses increase with age.

Pump Control

Pump capacity can be controlled by:
- On-off switching,
- Two-speed motor control,
- A control valve in the discharge of a pump,
- Variation in the speed of the pump and
- Stroke adjustment.

On-Off and Two-speed Control

The on-off type of pump control can be programmed with a timer or counter to deliver a known volume of fluid. The pump can deliver one full stroke and then stop until the next signal is received from the timer. A specified number of pump strokes can also be used. The flow can be smoothed by a pulsation dampener.

A two-speed motor can be used for simple pumping systems where the accurate control of pump pressure is not required. Most typical applications of this type of drive have appreciable pipe friction. They should not be used in systems with high static head and low friction.

Throttling Control

Continuously variable flow control at constant speed can be done by the automatic adjustment of the stroke length. The practical range of flow control by stroke adjustment is 10 to 100%. In some applications, reciprocating pumps combine the measuring and control functions and no independent feedback is used to represent flow. In other cases it is used as a final control element and flow detection is done with an independent sensor. Variable-speed control is often used in multiple-head pumps in which all the pumps are coupled mechanically. A control signal can be used to adjust all flows in the same proportion simultaneously.

The use of these methods of capacity control depends on the pump type such as centrifugal, rotary, or reciprocating as shown in Table 3-4.

Table 3-4 Pump Control Methods

Method of Control	Possible Types of Pumps
On-off switch	Centrifugal, rotary, or reciprocating
Throttling control valve	Centrifugal or rotary
Speed control	Centrifugal, rotary, or reciprocating
Stroke adjustment	Reciprocating

Reciprocating or Metering Pumps

Reciprocating pumps, such as the piston and diaphragm types, deliver a fixed volume of fluid per stroke. The control of these pumps is based on changing the stroke length, changing the stroke speed, or varying the interval between strokes. The discharge from these pumps is a pulsed flow, so they are not suited to control by throttling valves.

Rotary Pumps

The pump characteristics for rotary pumps, such as the gear, lobe, screw, or vane type, have a fairly constant capacity at constant speed with large changes in discharge pressure. The operation of rotary pumps with on-off control is similar to that of centrifugal pumps.

Intermittent feed flow can be obtained by opening an on-off valve via a signal from the cycle timer. A pressure-controller bypass allows normal pump flow to be maintained in the loop when the feed valve is closed. The on-off control in this case is applied to the fluid rather than to the pump motor.

Safety and Throttling Control

A safety relief valve is always used on a rotary pump to protect the system and pump casing from excessive pressure. The relief valve can discharge to pump suction or to the feed tank.

The output of a rotary pump can be continuously varied by using a pressure-controlled bypass in combination with a flow control valve. The bypass is used to accommodate changes in flow, since the total flow through the bypass plus the process is constant at constant pump speed.

The capacity of a rotary pump is proportional to its speed, neglecting any losses due to slippage. This flow can be controlled by speed-modulating devices. In this case no bypass is needed. The range is limited by the speed-control device, which for magnetic drives is approximately 4:1.

Centrifugal Pumps

Centrifugal pumps are the most common type of process pump. Capacity varies widely with changes in discharge pressure and the shape of the curve determines the type of control that may be used. At low flow rates there may be two flows that correspond to the same head, which causes control to be unstable in this region.

Pump Starters

A simple pump starter can use a three-position hand-off-automatic switch. Most pump controls include some safety overload (OL) contacts as shown in Figure 3-8. These overrides can be used to stop the pump if excessive pressure or vibration is detected. There is usually also contact for remote alarming. Most pump controls also include a reset button that must be pressed after a safety shutdown condition is cleared before the pump can be restarted.

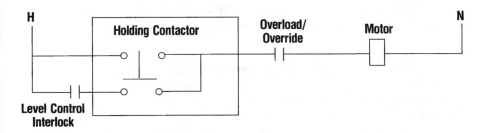

Fig. 3-8 On-off pump interlocks

When a group of starters is supplied from a common feeder, the starters should not be allowed to be overloaded by the inrush currents of simultaneously started units. A 30-second time delay can be used to prevent other pumps from starting until the delay has passed.

When two-speed pumps are used, an interlock can be used to guarantee that even if the operator starts the pump in high, it will operate for 30-seconds in low before advancing automatically to high. This makes the transition from off to high speed more gradual. Another interlock can be used to ensure that when the pump is switched from high to low, it will be off high speed operation for 30-seconds before low-speed drive is engaged. The use of tandem pumps allows a wide range of flows at high pump efficiency without serious loss in pressure head.

On-Off Pressure Control

A pressure switch can be used to start a spare pump to maintain pressure in a critical service when the operating pump fails. A low pressure switch is used to actuate the spare pump, which is piped in parallel with the first pump. Another function of the extra pump is to boost pressure. When the flow is reduced below a certain point, the pressure switch will start the parallel pump and close a bypass valve.

Throttling Control Valves

Throttling control can be achieved with control valves. The control valve is usually sized to pass the desired flow with a pressure drop equal to the system dynamic friction losses excluding the control valve but not less than 10 PSID (69 kPa) minimum. Pump flow is controlled by varying the pressure drop across the valve.

To avoid too low a flow, a bypass line can be used with a backpressure regulator. It is set at a pressure that will guarantee minimum flow as the pump is throttled toward zero flow. In a typical flow control loop, the range of the control valve will be about 20:1.

If a minimum flow bypass cannot be provided, then a safety interlock should be used. In this interlock a minimum flow switch is used to stop the pump on low flow.

Some energy is wasted through throttling which is part of the total loss of efficiency. The pump efficiency loss can be about 10%.

Pump Speed Throttling

In order to vary pump speeds with electric motors, it is generally necessary to use a variable-speed device in the power transmission train. This can consist of variable pulleys, gears, magnetic clutch, or hydraulic coupling. Variation of the pump speed produces a family of head-capacity curves, where the volume flow is proportional to speed if the impeller diameter is constant. The intersection of the system curve with the head curve determines the flow rate.

Speed throttling will not usually reduce the pump efficiency as much as valve throttling. This increases the total energy saving obtained from pump speed control. When the flow is reduced, instead of wasting the excess pump head of a pressure drop through a valve, the excess pump head is not introduced.

Variable-speed Pumps

In most friction systems the savings represented by variable-speed pumping increases with reduced pump loading. If, on a yearly average, the pumping system operates at not more than 80% of design capacity, the installation of variable-speed pumps can result in a payback period of approximately three years.

The savings resulting from variable-speed pumping can be calculated from the operating point on the pump curve. The relationship between the demand for flow and the power input required can be plotted. The difference between the curves for both control valve throttling and variable-speed pumping systems represents the savings potential. The savings due to variable-speed pumping increases as the flow rate drops off.

Variable-Speed Drive Efficiency

The overall system efficiency depends on the following factors:
- Pump efficiency (%),
- Motor efficiency (%),
- Variable-speed drive efficiency (%) and
- Efficiency of utilization (%).

Pump efficiency peaks at about 80% and drops to zero as flow decreases. Motor efficiency reaches about 95% at about 75% load. The efficiency of the variable-speed drive depends on the type of drive used. Each design has a different efficiency. The maximum varies from 90 to 95% at the top speed and can vary from 40 to 90% at half speed. Compared to valve throttling, the energy savings are less with electromechanical or slip control drives than with solid-state electrical drives.

The efficiency of utilization depends on the specific water distribution system design. If the piping distribution system uses three-way valves, the actual flow is much higher than the required flow, and the head required to transport the actual flow is also greater than the required head. The efficiency of utilization is the ratio of the required and actual products.

$$E = \frac{Q_{required}}{Q_{actual}} \quad \frac{H_{required}}{H_{actual}} \quad 100$$

The actual value can be accurately obtained only by testing.

Types of Variable-Speed Drives

Variable-speed pumps usually operate within the speed range of 50 to 100% but the drive should be able to function at speeds down to zero without damage or overheating.

Variable-speed pump drives are either electrical or electromechanical. The electrical types include two-speed, direct current, variable voltage, variable frequency (pulse width modulated), and wound rotor regenerative.

All of these types depend on altering the characteristics of the electrical power to the motor in order to change its speed. The electromechanical drives use a speed-changing unit that is located between the motor and the pump. These units include eddy-current couplings, hydraulic or hydroviscous drives, and V-belt drives.

Electromechanical drives are generally less efficient than the electrical drives. The variable voltage electric drive using high-slip motors is an exception, since it is less efficient than the mechanical drives. Table 3-5 provides a summary of the various drive types.

Table 3-5 Variable-speed Drives

Electric Motors
Wide speed range, high output power at variable speed, accurate speed control.

Eddy current or magnet couplings
High output power at variable speed, accurate speed control.

Continuously variable mechanical drives
Small to moderate load with narrow speed range.

Hydraulic drives
Shock loads, overloads, accurate speed control.

The variable frequency drives are among the most efficient units. As the frequency sent to a squirrel-cage induction motor changes from 60 to 40 Hz, the pump speed is correspondingly reduced from 100% to 60%. These drives are typically available in sizes up to 250 HP (185 kW), with special units being available up to 2000 HP (1491 kW).

In a conventional wound rotor unit, the motor speed is altered by varying the resistance in the rotor. This wastes energy and is inefficient. In the regenerative units, the resistors are eliminated, and inverter control is used which is much more efficient. Sizes vary from 25 to 10,000 HP (20 to 7,455 KW). A wound rotor motor is more expensive than induction types, and the brushes used in the regenerative design require replacement every four years.

DC variable-speed drives using wound rotors provide stepless speed control at high efficiency and modest cost. Since the armature rotates at right angles to the magnetic field established by the field coils, the rotation is easily reversed by reversing polarity. One advantage of these drives is the ability to maintain a constant speed while the load varies. Units are available in sizes of 1 to 500 HP (0.75 to 373 kW).

Variable voltage drives use a special high-slip induction motor to reduce the speed as the voltage is reduced. This type of drive has low efficiency with a high level of heat generation. It is not recommended for centrifugal pumps.

Two-speed motors can be used in simple pumping systems where accurate control of pump pressure is not required. These motors should not be used in systems with high static head and low friction. In higher friction systems they offer moderate efficiency in a low cost alternative to variable-speed pumps. Typical motor speeds include:

- 1750/1150 rpm (29 and 19 r/s),
- 1750/850 rpm (29 and 14 r/s),
- 1150/850 rpm (19 and 14 r/s) and
- 3500/1750 rpm (58 and 29 r/s).

In an eddy-current coupling drive, the motor rotates a drum or rotating ring. Inside the drum or ring is an electromagnetic rotor assembly that is connected to the pump. Speed-control depends on the amount of magnetic flux between the drum and rotor assemblies. The slip or speed difference between the motor and pump increases and decreases with the flux density in the coupling. Because it is a slip-type, variable-speed device, the eddy-current coupling does not have as high an efficiency as the variable frequency, direct current, and wound rotor regenerative types of variable-speed drives. It is rugged and used widely in large pump applications.

Fluid couplings consist of two coupling halves, one is driven by the motor and the other is connected to the centrifugal pump. The pump shaft speed increases with the amount of oil supplied to the coupling. It is a low cost design and is used where efficiency is not important.

Mechanical drives are used on pumps smaller than 100 HP (75 kW). Rubber belts or metal chains are used to adjust the speed within a range of 6:1. They suffer from relatively low reliability and efficiency.

Multiple Pumps

Multiple pumps in parallel can eliminate overpressure at low flows. When constant-speed pumps are used in parallel, the second pump can be started and stopped automatically based on the flow. An adjustable dead band is provided with a flow switch. This dead band starts the second pump when the flow rises above the upper end, but will not stop until it drops below the lower end. This dead band action prevents excessive cycling.

Booster Pumps

With most friction loads, series pumping can greatly reduce the overpressure at low loads. Multiple pumps in series are preferred for operating costs, but a single two-speed pump represents a lower capital investment. If constant-speed pumps are used, the booster pump can be started and stopped automatically with pressure. An adjustable dead band is provided with a pressure switch.

Water Hammer

When a valve opening is suddenly reduced in a moving water column, a pressure wave will travel in the opposite direction to the water flow. When the pressure wave reaches a solid surface such as an elbow or tee, it is reflected back to the valve. If the valve has closed, a noise sounding like a series of hammer blows results. Methods of preventing water hammer include using low velocities, valves with slow closure rates or slow-closing bypass valves around fast-closing valves.

When the cause of water hammer cannot be corrected, it can be treated by adding air chambers, accumulators, or surge tanks. Surge suppressors, such as positively controlled relief valves can also be used. Vacuum breakers can be used to admit air and cushion the shock resulting from the sudden opening or closing in a split stream system.

Optimizing Pump Controls

The potential for optimization depends on the capabilities of the equipment and piping configuration. One type of optimization of a pump system consists of a variable-speed and a constant-speed pump. A pressure controller maintains a minimum pressure difference between the supply and return from each area. Thus, no user can be denied more hot water if its control valve opens further. If demand drops, the pressure controller slows down the variable-speed pump to keep the differential at the areas from rising above the minimum. The differential pressure transducer should measure the head loss not only of the load but also of the control valve, hand valves, and piping.

When the variable-speed pump reaches its maximum speed, a high pressure switch starts pump 2. When the load drops down to the setpoint of a low flow switch, the second pump is stopped. Note that the pumps are started by pressure but stopped by flow controls.

Low flow switches are also used as safety devices to protect the pumps from overheating or cavitation, which can occur when the pump capacity is very low. The pump flow rate should be greater than 1 GPM (3.78 l/m) for each pump break horsepower. If the flow rate drops below this limit, the pumps are stopped by the low flow switches.

Variable-speed pumping cannot completely replace the use of control valves, but variable-speed pumps are a more energy-efficient method of fluid transportation and distribution than constant-speed pumps used in combination with control valves. If the fluid is being distributed in a mostly friction system and the average loading is under 80% of pump capacity, the installation of variable-speed pumps can reduce operating costs.

Valve Control Optimization

The optimum pump discharge pressure is that which will keep the most open user valve at a 90% opening. As the pressure rises, the user valves close and as it drops, they open. Opening the most open valve to 90% causes the others to be opened also. This keeps the pump discharge pressure and the pumping energy at a minimum.

A valve position controller is used to open the valves. It minimizes the valve pressure drops and reduces valve cycling and maintenance. This is because cycling is more likely to occur when the valve is almost closed, and maintenance is high when the pressure drop is high.

The pressure controller setpoint is changed slowly since the valve position controller has integral action only, and its integral time is set to be about 10 times that of the pressure controller. In order to prevent reset windup when the pressure controller is switched from cascade to automatic, the valve position controller should have external feedback.

In a pump station of two variable-speed pumps, their speed is set by a pressure controller. When only one pump is in operation and the pressure controller output reaches 100%, a high pressure switch is actuated and the second pump is started.

When both pumps are in operation and the flow drops to 90% of the capacity of one pump, the second pump is stopped provided that a time delay has occurred. The time delay insures that the pump is not started and stopped too often.

As the demand for water rises, the speed of the operating pump increases until 100% is reached, then a high pressure switch starts the second pump. In order to eliminate a temporary surge in pressure, a signal generator is used for the required speed of the second pump.

Once both pumps are operating, the next control phase is to stop the second pump when the load drops where it can be met by a single pump. This is controlled by a low flow switch, which is set at 90% of the capacity of one pump.

Another type of pump control uses the pressure drop through an in-line orifice for pump speed control. The orifice drop can be related to the pipeline pressure losses by multiplying the drop by a set resistance ratio. The terminal pressure can then be maintained under variable load conditions by adjusting the pump speed.

Blowdown

A goal of optimization is to minimize blowdown as much as possible without causing excessive sludge or scale buildup on the inside surfaces of the boiler tubes. The benefits of this optimization include the reduction in the need for makeup water and treatment chemicals and the reduction in heat loss as hot water is discharged.

Blowdown can be optimized by automatically controlling the chloride and conductance of the boiler water. A neutralized conductivity setpoint of around 2500 micrombos can be controlled automatically to about +/- 100 micrombos.

The rate of blowdown accelerates as the boiler water conductivity setpoint is lowered. A reduction of about 20% can result from converting the blowdown controls from manual to automatic.

Overall boiler efficiency can also be increased if the heat content of the hot condensate is returned to the boiler. Pumping water at high temperatures is difficult and it is best to use pumpless condensate return systems. In operation of such a system, the steam pressure is used to push back the condensate into a de-aeration tank. This approach eliminates not only the maintenance and operating cost of the pump but also any flash and heat losses resulting in the return of more condensate at a higher temperature.

The performance of the steam and condensate piping system can also be improved if steam flows are metered. This data is helpful in accountability calculations as well as in locating problems such as insufficient thermal insulation or leaking.

Load Allocation

Load allocation between several boilers involves the distribution of the total demand in the most efficient and optimized manner. This type of optimization should reduce the steam production cost to a minimum.

A computer-based energy management system can operate either in an advisory or in a closed-loop mode. A closed-loop system can automatically enforce the load allocation without the need for operator involvement. An advisory system provides instructions to the operator and leaves the implementation up to the operator. In a simple load allocation system, only the starting and stopping of the boilers is optimized. If the load is increasing, the most efficient idle boiler will be started and when the load is dropping, the least efficient one is stopped.

More sophisticated systems can optimize the load distribution between operating boilers. Software packages for continuous load balancing are available. The computer is used to calculate the real-time efficiency of each boiler and this is used to calculate the incremental steam cost for the next load change for each boiler. When the load increases, the incremental increase is sent to the setpoint of the most cost-effective boiler. If the load decreases, the incremental decrease is sent to the least cost-effective boiler.

Chapter 4
Heat Pump and Chiller Optimization

Heat pumps move heat from a lower to a higher temperature level. They do not create heat. This is similar to a water pump moving water from a lower to a higher elevation.

The energy required to do this depends not so much on the amount of water to be moved but by the elevation to which it is lifted. In a similar way, the required energy for a heat pump depends not only on the amount of cooling to be done but also on the temperature elevation. The optimization of heat pumping involves the reduction of this temperature difference. Each degree F reduction in the temperature difference will lower operating costs by about 1.5%, for each degree C this is about 2.7%.

Heat Pump Thermodynamics

A heat pump removes an amount of heat (QL) from the chilled refrigerant and invests an amount of work (W) to deliver a quantity of heat (QH).

In the ideal temperature-entropy cycle for a refrigerator, there are two isothermal and two adiabatic processes. (Figure 4-1) In a refrigeration machine, as it goes through the expansion valve, the high pressure subcooled refrigerant liquid becomes a low pressure, liquid-vapor mixture. As it goes through the evaporator, this becomes a superheated low-pressure vapor stream. Then, in the compressor, the pressure of the refrigerant vapor is increased. Finally, this vapor is condensed at constant pressure. The adiabatic processes take place in the expansion valve and the compressor. The isothermal processes take place though the evaporator and the condenser. The isothermal processes in this cycle are also isobaric since they occur at constant pressure.

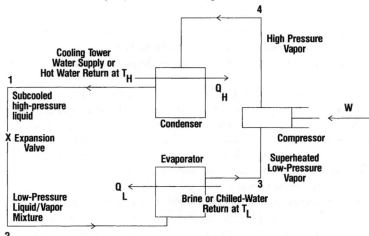

Figure 4-1 Refrigeration cycle with two isothermal and two isentropic processes

The efficiency of a refrigerator is defined as the ratio between the heat removed from the process (QL) and the work required to achieve this heat removal (W).

efficiency = $\dfrac{QL}{W} = \dfrac{TL}{TH - TL}$

The chiller efficiency is greater than 100% and it is usually called the coefficient of performance (COP). When a chiller requires 1.0 kWh (3412 BTU/h) to provide a ton of refrigeration (12,000 BTU/h), its coefficient of performance is said to be 3.5. This means that each unit of energy introduced at the compressor will pump 3.5 units of heat energy to the cooling tower water. Typically, chillers operate with COPs in the range of 2.5 to 3.5. Optimization can double the COP by increasing Tl or decreasing Th.

The unit most often used for refrigeration loads is the ton.

Several types of tons are used with the following differences:

Standard ton	**200 BTU/minute (3520 W)**
British ton	**237.6 BTU/minute (4182 W)**
European ton (Frigorie)	**50 BTU/minute (880 W)**

Refrigerants

The fluid that carries the heat from a low to a high temperature level is called a refrigerant. Table 4-1 lists some common refrigerants.

Most working fluids can be used with reciprocating compressors, but only the last four fluids with their high volume-to-mass ratios and low compressor discharge pressures are used with rotary or centrifugal machines.

Operation under vacuum in any part of the cycle can introduce sealing problems. Very high condensing pressures are also avoided because of structural strength requirements. When low temperatures are required, ethane is used with its high system pressure.

If the latent heat of the refrigerant is high, more heat can be carried by the same amount of working fluid. This allows the equipment to be smaller. Although, ammonia has some undesirable characteristics, it has been used for this reason.

An important consideration is safety; it is desirable to use nontoxic, nonirritating, nonflammable refrigerants. The working fluid should also be compatible with compressor lubricating oil. Corrosive refrigerants are undesirable because of the increased maintenance. Freon-12 is the most suitable except for its problems with ozone depletion and the fact that it is being phased out.

In many applications the refrigerant evaporator is not used directly to cool the process. The evaporator cools a circulated fluid, which is then piped for cooling.

Brine is also used at temperatures below the freezing point of water can be used as a coolant. Strong brines may not freeze, but unless they are not true solutions, plugging of evaporator tubes can occur. Sodium brines can be used at 0°F

(-18°C) and calcium brines can be used for services down to -45 F (-43 C). Brines are corrosive and will initiate galvanic corrosion between dissimilar metals.

Table 4-1 Refrigerant Characteristics

Refrigerant (Low to high boiling points)	Compressor Type	Toxic Flammable Irritating	Mixes Compatible Lubricating Oil	Features
Ethane	R	T, F	No	Low temp. service
Propane	R	T, F	No	
Freon-22	R	No	OF	Low temp. service
Ammonia	R	T, F	No	High efficiency refrigerant
Freon-12	R	No	Yes	Most recommended
Methyl chloride	R	A	Yes	Expansion valve may freeze if water present
Sulphur dioxide	R	T, I	No	E, L, H
Freon-21	RO	No	Yes	E, L, H
Ethyl chloride	RO	F, I	No	E, L, H
Freon-11	C	No	Yes	E, L, H
Dichloromethane	C	No	Yes	E, L, H

Key:
C = Centrifugal
E = Evaporator under vacuum
H = High volume-to-mass ratio across compressor
L = Low compressor discharge pressure
R = Reciprocating
RO = Rotary
OF = Oil floats on it at low temperature
A = Anesthetic

Refrigerator Control

In the direct expansion type of refrigeration control, a pressure-reducing valve maintains constant evaporator pressure. The pressure setting depends on the load, and these controls are used for constant loads. The correct setting is found by adjusting the pressure-control valve until the frost stops just at the end of the evaporator indicating the presence of liquid refrigerant up to that point. When the load increases, more of the refrigerant vaporizes. This causes low efficiency in the unit since it is starved.

The thermostatic expansion type of control does not maintain evaporator pressure, it controls the superheat of the evaporated vapors. This design guarantees the presence of liquid refrigerant at the end of the evaporator under all load conditions. In the standard type of control, the input pressure and output temperature of the evaporator are detected.

Provided that the evaporator pressure drop is low, these measurements of the saturation pressure and the temperature of the refrigerant are an indication of superheat. The desired superheat is set by a spring in the valve operator, which along with the saturation pressure in the evaporator opposes the opening of the valve.

When the load increases, it causes the evaporator outlet temperature to rise. This results in a rise in the temperature bulb pressure, which opens the control valve. The increased saturation pressure balances against the increased bulb pressure for a new equilibrium.

A cooling water regulating valve may also be used. (Figure 4-2) This valve maintains the condenser pressure constant and conserves cooling water. At low condenser pressure, such as when the compressor is off, the valve closes. It opens when the compressor is started and its discharge pressure reaches the setting of the valve. The water valve opening follows the load, and continues to open at higher loads to maintain the condenser pressure constant.

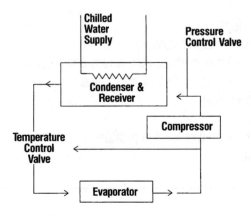

Figure 4-2 Expansion valve system with water regulating valve

At very low temperatures a small change in refrigerant vapor pressure corresponds to a fairly large change in temperature. This allows a thermal bulb to indirectly detect the saturation pressure in the evaporator. A more sensitive measurement can be obtained using this technique with a differential temperature expansion valve as shown in Figure 4-3.

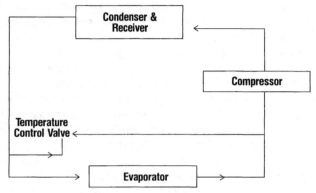

Figure 4-3 Differential temperature expansion valve

On-Off Control

A small refrigeration unit can include the following:
- Superheat control valve,
- Low-pressure-drop evaporator,
- Reciprocating on-off compressor and
- Air-cooled condenser.

A high-temperature switch acts as the main control device. When the temperature of the refrigerated water drops below a preset value, usually 38°F, or 3.3° C, the refrigeration unit is turned off. Then, when it rises to a higher level, usually 42°F, or 5.6°C, it is restarted. This on-off control is done with a solenoid valve which is controlled by the temperature switch. The closing of the valve causes the compressor suction pressure to decrease until it reaches the setpoint of a low-pressure switch which stops the compressor.

When the unit is running, an expansion valve maintains the refrigerant superheat constant. Safety interlocks to protect the equipment include turning off the compressor if the fan motor stops or if the compressor discharge pressure becomes too high.

This type of operation is called two-stage unloading. Varying the cooling capacity of the unit beyond this can be done by adding more stages or steps.

Commercial Chillers

Supermarkets, warehouses, and other facilities often have a large number of loads or zones which are served by an evaporator coil. There are a number of positive-displacement refrigerant compressors which are turned on or off to follow that load.

The optimization of these systems involves minimizing the cost of operation by meeting the cooling loads with a minimum of compressor horsepower. The compressors must be operated in such a way that the total run-time of each machine is about the same.

Each cooling zone may be a separate room, freezer, or other cooled area with its own temperature control. Each zone is cooled by an evaporator coil, served by liquid refrigerant. An expansion valve controls the superheat of the evaporated vapors by sensing the pressure and temperature of the vapors.

The expansion valve allows as much refrigerant liquid as required to keep the evaporator superheat constant. This is the temperature above the boiling point at the pressure where the refrigerant is being evaporated.

A balancing valve is provided for each zone. These are located in the refrigerant vapor line leaving the zone. When all zones are at the same temperature, the balancing valves are full open, which minimizes the pressure drop on the suction of the compressors and maximizes their efficiency.

If some of the zone temperatures are at a higher temperature than others, the balancing valves on the higher temperature zones is throttled to increase the vaporization pressure and the boiling temperature of the refrigerant.

If the zones are operated near freezing temperature (below 35°F, or 2°C), defrosting control is needed because of the ice buildup on the evaporator coils. The defrosters are controlled by a coil surface temperature detector switch in each zone. When the coil surface temperature is below the low limit for a set time period, the switch shuts off the refrigerant liquid supply by closing a solenoid valve. Then, when all the refrigerant has been vaporized, a coil defroster heater is turned on for a short time to remove the ice. During this defrosting cycle the zone temperature controller is disabled by setting its measurement and setpoint to be equal.

Each additional compressor unit is turned on as the suction pressure to the compressor station drops to a lower level. This is inefficient, since these lower levels of suction pressure requires additional compressor horsepower.

An optimized control system eliminates this wasted energy by using high and low pressure switches for turning additional compressors on or turning unnecessary compressors off. The pressure switches trip a time delay and if the pressure switch is actuated for this time period, it increases or decreases the horsepower level of the compressors. Two time delays are used. The response to a load increase is in minutes, while the response to a load decrease is in seconds, which serves to maximize the energy savings.

Additional optimization is achieved by increasing the control gap between the pressure switches as high as the loads will allow. This continuous floating of the controlling suction pressure gap uses feedforward action based on the zone temperature control, which is a reverse-acting, integral-only controller.

As zone temperatures rises above the zone thermostat setpoint, the output signal of the thermostat drops. The lowest of the thermostat outputs is selected which identifies the zone with the highest cooling load. If this load is satisfied, all the others will be satisfied.

The setpoint of the high pressure switch is obtained by summing 90% of the suction pressure of the compressor station with the output signal of the highest load thermostat. The setpoint of the low pressure switch is 95% of this setpoint.

Energy consumption is minimized by maximizing these setpoints, since when the zone temperatures are at setpoint or below, the suction pressure of the compressor station is increased and the amount of work the compressors must do is less. The total energy savings from this type of control system is about 20%.

Larger Chillers

Refrigeration units in the 500 ton (1760 kw) or larger range are capable of continuous load adjustment. They may also use an economizer expansion valve system or hot gas bypass to increase rangeability. Load changes are reflected by the temperature of the returning refrigerated water.

The load decreases, the return water temperature will drop and a temperature controller will gradually close the suction damper or the prerotation vane of the compressor. Throttling the suction vane allows a 10:1 turndown ratio. If the load drops below this ratio, the hot gas bypass system is activated.

The hot gas bypass is controlled by another temperature controller. It keeps the constant-speed compressor from surging. When the load drops to a level approaching surge conditions, the bypass valve is opened. The difference between chilled water supply and return temperatures is used to indicate the load conditions. Throttling action is accomplished by a proportional controller.

As the hot gas bypass allows the chiller to operate at low loads without going into surge, there is an increase in operating costs. This is the work wasted by the compressor through the hot gas bypass valve. An optimized control system can reduce this waste by using a variable speed compressor.

Instead of controlling the hot gas bypass based on return water temperature, it can be controlled on the basis of prerotation vane opening. Opening the hot gas bypass when the prerotation vane has closed to some fixed point disregards condensing temperature. This causes the hot gas bypass to open sooner and wider extent than necessary. An optimized control will recognize cooling water temperature variations and open the hot gas bypass valve only when needed.

Economizer

An economizer can increase the efficiency of operation by 10%. There are savings in compressor power consumption, reduction of condenser and evaporator surfaces, and other effects. The economizer is a two-stage expansion valve with condensate collection chambers. The savings are a result of vaporization in the lower chamber from precooling the liquid that enters the evaporator. At the same time, the vapors are superheated that are sent to the compressor second stage.

Safety Interlocks

An interlock system prevents the compressor motor from being started if any of the following conditions exist and stops the compressor if any except the first condition occurs:

- Open suction vane, detected by limit switch;
- Low water temperature, near freezing point, sensed by low temperature switch;
- Low water flow, sensed by low flow switch;
- Low evaporator temperature, near freezing point, detected by low temperature switch;
- High compressor discharge pressure, sensed by high pressure switch;
- High motor bearing or winding temperature, detected by high temperature switch and
- Low oil pressure.

Another interlock system guarantees that the following are operating upon starting of the compressor: water pump, oil pump and water to oil cooler. The suction vane usually has an interlock to be sure it is completely closed when the compressor stops.

In a typical cooling system, the cooling load is carried by the chilled water to the evaporator, where it is transferred to the Freon or other refrigerant. The refrigerant takes the heat to the condenser where it is passed on to the cooling tower water and is finally given up to the ambient air.

Cooling Costs

This type of heat pump system involves four heat transfer substances (chilled water, refrigerant, cooling tower water, air) and four heat exchanger devices (heat exchanger, evaporator, condenser, cooling tower). The system operating cost is the cost of circulating the four heat transfer substances.

In a conventional, unoptimized system each of these four systems operate independently in an uncoordinated manner. In a conventional system, the transportation devices operate at constant speeds and they introduce more energy than is needed for the circulation of refrigerant, air, or water. This results in a waste of energy.

Load-Following

Load-following is the key to optimization which eliminates this waste. The system must be operated as a coordinated single process. The goal of control is to maintain the cost of operation at a minimum.

The controlled variables are the supply and return temperatures of chilled and cooling tower waters while the manipulated variables are the flow rates of chilled water, refrigerant, cooling tower water, and air.

Water temperatures can be allowed to float in response to load and ambient temperature variations. This eliminates the waste needed to keep them at arbitrary fixed values and reduces chiller operating costs.

In a total refrigeration system, four control loops need to be configured. The controlled or measured variables are the four water temperatures. The four manipulated variables are the motors for the transportation devices. Table 4-2 shows the controlled and the manipulated variables in these loops as well as the optimization criteria for the setpoints.

Table 4-2 Control and Optimization – Chiller Control Loops

Controlled Variable	Manipulated Variable	Optimization Technique Setpoint for Temperature Control
Cooling tower water supply temperature	Fan operation	Minimize compressor, fan and cooling tower water pump costs TIC = wet-bulb temperature + optimium range of cooling tower water
Cooling tower water return temperature	Cooling tower water pump	Find optimum cooling tower range setpoint = supply temperature + range
Chilled water supply temperature	Chiller compressor	Setpoint = maximum supply temperature which meets all loads
Chilled water return temperature	Chilled water pump	Setpoint = maximum return temperature which meets all loads

The costs of operating a cooling system can be divided up as follows: fans 10-15%, cooling tower water pump 15-20%, compressor 50-60% and chilled water pump 15-20%.

Fan costs increase in geographic regions with warmer weather. Longer water lines with more friction cause an increase in pump costs. Compressor costs become lower as the maximum allowable chilled water temperature rises.

In any particular installation, the goal of optimization is to find the minimum chilled water and cooling tower water temperatures which will result in meeting the cooling needs of the installation at minimum cost.

Cooling Towers

In a cooling tower the heat and mass transfer processes are combined to cool water. The mass transfer from evaporation consumes water. This loss is about 5% of the water needed for an equivalent once-through cooling by a stream of water. Cooling towers are capable of cooling within 5 to 10°F (2.8 to 5.6°C) of the ambient wet-bulb temperature. The larger the cooling tower for a given set of water and airflow rates, the smaller this temperature difference will be.

The psychrometric chart can be used to describe the properties of air in the cooling tower. Ambient air can enter in various conditions, which are points on the chart. After transferring heat and absorbing mass from the evaporating water, the air leaves in a saturated condition. The total heat transferred depends on the difference between the entering and leaving enthalpies.

Any component along the dry-bulb temperature axis represents the heat transfer component (sensible air heating), and any component along the vapor axis represents the evaporative mass transfer component (latent air heating). The water is sensibly heated by the air and cooled by evaporation. As long as the entering air is not saturated, it is possible to cool the water with air that is warmer than the water.

The basic cooling tower designs are described in Table 4-3.

Table 4-3 Types of Cooling Towers

Type	Fan	Features
Forced-Draft Crossflow	Bladed, top fan	Hot water in sprays on fill Air out on top, louvered bottom with cold water outlet
Forced-Draft Counterflow	Centrifugal, bottom fan	Hot water in sprays on fill Air forced out by bottom fan
Forced-Draft Crossflow	Bladed, side fan	Hot water in flows on fill Air out on side, louvers on side and bottom, cold water out on bottom
Induced-Draft Crossflow	Bladed, top fan	Hot water in flows on fill Air out on top, louvers on bottom with cold water out Single and double air entry

Cooling tower characteristic curves are based on the following relationship.

$$\frac{W^n}{A}$$

Where

W = **Mass water flow rate, lb per hour**

A = **Airflow rate, lb dry air per hour**

The coefficient n varies from -0.35 to -1.1 and has an average value of -0.6.

The cost of a cooling tower depends mainly on the required water flow rate. It is influenced by the temperature difference, range, and wet-bulb temperature. The cost tends to increase with range and decrease with a rise in difference and wet-bulb temperatures. The cost of operation reaches the cost of the initial investment in 5 to 10 years, depending on energy costs.

Controls

The cooling tower is an air-to-water heat exchanger with controlled variables being the supply and return water temperatures, and manipulated variables, the air and water flow rates. Manipulation of the flow rates may be continuous, using variable speed fans and pumps, or incremental, using the cycling of single- or multiple-speed units. If the temperature of the water supplied by the cooling tower is controlled by modulating the fans, a change in cooling load means a change in the operating level of the fans.

If the fans are single-speed units, the control must cycle them on or off as a function of the load. This may be done by a simple sequencer or a more flexible digital control. In most areas, the need to operate all fans at full speed occurs for only a few thousand hours every year. Fans running at half speed consume about a seventh of the design horsepower but produce more than half of the design air rate or cooling effect. Most of the time the entering air temperature is less than the design rating, so two-speed motors can be used for minimizing fan operating costs.

Another reason for considering the use of two-speed fans is to have the ability to lower the associated noise level at night or during periods of low load. Switching to low speed will usually lower the noise level by approximately 15 dB.

Interlocks

Most fan controls include safety overrides in addition to overload contacts. These overrides can stop the fan if fire, smoke, or excessive vibration are detected. They also provide a contact for remote alarming. Most fan controls include a reset button that must be pressed after a safety shutdown condition is cleared before the fan can be restarted.

When several starters are supplied from a common feeder, it may be necessary to limit the inrush currents of simultaneously started units. A 30-second time delay can be used to prevent some fans from starting until that time has passed. Large fans should be protected from overheating caused by too frequent starting and stopping. This can be provided with a 0 to 30-minute time delay which assures that the fan will not be cycled at a frequency faster than the delay time setting.

If two-speed fans are used, additional interlocks may be needed. An interlock can guarantee that the fan will operate for 20 to 30-seconds in low before advancing to high. This will make the transition from off to high speed more gradual.

Another interlock can assure that when the fan is switched from high to low, it will be off the high speed for 20 to 30-seconds before the low-speed drive is engaged. This will give time for the fan to slow before the low speed is engaged.

If the fan can be operated in reverse, other interlocks may be used. One interlock can guarantee that the fan is off for several minutes before a change in the direction of rotation can take place. This gives time for the fan to come to rest before making a change in direction. Another interlock can guarantee that the fan can operate only at slow speed in the reverse mode and that reverse operation cannot last more than 20 to 30-minutes. This interlock is used because in reverse ice tends to build up on the blades in winter.

Evaporative Condensers

Another cooling tower type is the evaporative condenser. Here, induced air and sprayed water flow downward, and the condensate vapors travel upward in multipass condenser tubes. A pressure controller modulates the amount of cooling to match the load. The continuous water spray on the tube surfaces improves the heat transfer and maintains tube temperatures.

The temperature difference in these cooling towers is about 20°F (11°C). The wet-bulb temperature of the cooling air can be varied by the recirculation of humid or saturated air. This results in condensing temperatures that are lower than those in dry-air-cooled condensers.

The higher heat capacity and lower resistance to heat transfer allows these units to be more efficient than dry-air coolers. The turndown capability of these units is also high, since the fresh airflow can be reduced to zero. Both the exhaust and the fresh air dampers close while the recirculation damper simultaneously opens to maintain a constant internal airflow rate. As more humid air is recirculated, the internal wet-bulb temperature rises and the heat flow drops off.

Tower Optimization

The goal in optimization is to minimize the fan and pump operating costs. The cost of fan operation can be reduced by allowing the cooling tower temperature to rise. At some point this will not produce a low enough temperature for the load.

As the cooling temperature difference is increased, the temperature controller will open its coolant valve and more water must be pumped. The pumping costs rise and the fan costs tend to drop. The optimum setpoint corresponds to the minimum point on this curve. The corresponding range value for the temperature can also come from this curve.

Supply Temperature Optimization

An optimization control loop is required in order to maintain the cooling tower water supply continuously at an economical minimum temperature. This minimum temperature is a function of the wet-bulb temperature of the atmospheric air. The cooling tower cannot provide a water temperature that is as low as the ambient wet-bulb, but it can approach it.

As the temperature difference increases, the cost of operating the cooling tower fans drops and the cost of pumping increases. The optimum point is the one that will allow operation at an overall minimum cost. This point becomes the setpoint of a temperature difference controller. If the fans are single or two-speed units, the temperature difference controller will incrementally start and stop the fan units in order to maintain the optimum approach. If the cooling tower fans are centrifugal units or if the blade pitch is variable, continuous throttling is used.

Return Temperature Optimization

The control of return water temperature optimization requires a temperature difference range controller with an optimized setpoint. The setpoint is the optimum temperature range value corresponding to that cooling tower.

The range controller throttles the water circulation rate in order to keep the range at the optimum value. The output signals of the range controller and load controller are sent to a high signal selector.

This satisfies the needs of the users for priority over range control. The user demand is established by a pressure difference controller. If its output is higher than that of the range controller, it will be used for setting the pump speed.

A load-following optimization loop allows all cooling water loads to be satisfied while the range is being optimized. This is done by selecting the most-open cooling water valve and comparing that signal with the 90% setpoint of the valve position controller.

If even the most-open valve is less than 90% open, the setpoint of the pressure difference control is decreased. If the valve opening exceeds 90%, the setpoint is increased. This allows all users to obtain more cooling by opening their supply valves.

Since all cooling water valves are opened as the water temperature difference across the load is minimized, valve cycling is reduced and pumping costs are lowered. The reduction in pumping costs results from opening all of the cooling water valves, which lowers the pressure drop. Valve cycling is reduced when the valve opening is moved away from the unstable area near the closed position.

The control system must be stable so it is necessary to use an integral-only controller for valve position control. The integral time is 10X that of the integral setting of the load controller. This will allow the optimization loop to be stable when the valve opening signal is either cycling or noisy.

A high limit setting on the output of the valve position control assures it will not move the cooling water pressure to unsafe or undesirable levels. Since this limit can block the valve control output from changing the pump speed, it is necessary to protect against reset windup. This is done with external feedback signal.

This type of optimization system can also point out system limitations. If the load controller is in control most of the time, the control valves and water pipes may be undersized. If the same valve is always selected as the most open, this means that the water supply to that valve is undersized.

These problems can be corrected by adding local booster pumps or by replacing undersized valves and pipes. The control should then return to the range controller. The load controller should act more like a safety override that becomes active under emergency conditions allowing no load to run low of cooling water.

Safety Controls

The minimum basin level is usually maintained with float-type level control valves. The maximum basin level is maintained by overflow nozzles sized to handle the total system flow. In some systems safety alarms include high/low level alarm switches. The safety of the fan and fan drive may include torque and vibration detectors. Low-temperature alarm switches are often used as warning devices for signaling the failure of freeze protection controls.

Flow Balancing

When a number of cooling tower cells are used, the water flows to the cells should be balanced as a function of the operation of the associated fans. The water flows to cells that have high speed fans should be controlled at equally high rates. Cells with fans operating at low speeds should receive water at low flow rates. Cells with their fans off should be supplied with water at minimum flow rates.

Typically, the water flow rate range is 2 to 5 GPM per ton when the fan is at full speed. A 20% change in water flow rate affect the cooling tower temperature difference by about 2°F.

Water balancing is often done manually, but it can also be done automatically. The total flow is used as the setpoint of a ratio flow controller for each cooling tower. When the ratio settings are the same, the total flow should be equally distributed. The ratio settings can be changed automatically to follow changes in fan speeds. The total of the ratio settings should always equal 1. If one ratio setting is changed, all the others must be modified.

This type of control system cannot only distribute the returning water between the cells but it can do it at minimum cost. The cost involved is the pumping cost. It will be a minimum when the pressure drop through the control valves is minimum.

This is done with valve position control. As long as the most open valve is not full open, the valve position controller adds a small positive bias to all the setpoint signals of the flow ratio controllers. This moves the valves open until the most open valve reaches a 95% opening and allow flow distribution at a minimum cost in pumping energy.

Manual balancing often results in all balancing valves being throttled with the same water flow when the fan is on or off. These conditions will increase operating costs, and this represents a major part of the savings from automatic balancing.

Winter Operation

In subfreezing weather any electrically heated tower basins should have emergency draining or bypass circulation. In a bypass system a thermostat detects the outlet temperature of the cooling tower water. When it drops below 40°F (4.4°C), the bypass circulation, which is located in a protected indoor area, is started. The circulation is stopped when the water temperature rises above 45°F (7.2°C). The thermostat is wired to open the bypass loop valves using solenoid control. It also starts a small circulating pump.

In very cold regions a main pump bypass is used. When the main pump is started, this bypass will not allow the water to be sent to the top of the tower until its temperature is above 70°F (21°C). When the water reaches this temperature, the bypass is closed since the water is warm enough to be sent to the top of the tower without danger of freezing. As the water temperature rises, the fans are turned on to provide added cooling.

When cooling towers are operated at freezing temperatures, induced draft fans need to be reversed periodically to de-ice the air intakes. The need for de-icing can be determined by visual observation or closed-circuit TV inspection. Another method is to compare the load and ambient conditions based on the operating history. Closed cooling towers can operate on antifreeze or be provided with supplemental heat and tracing.

Since blowdown involves the discharge of water to control the amount of impurities in the circulating water, there will be a water loss of about 1–3% of the circulating water rate. This loss depends on the amount of circulating water and the concentration of dissolved natural solids and chemicals added for protection against corrosion or buildup of scale.

Blowdown must meet the water quality standards for the accepting stream and must not be unreasonably expensive. It is also possible to make the water system a closed-cycle system that needs no blowdown with ion-exchange treatment.

Evaporator Optimization

Optimization of the evaporator section depends on making the compressor and chilled water pump costs a minimum when both the chilled water supply and return temperatures are at maximum. This holds for all load conditions, since the amount of work which the chiller compressor has to do is reduced as the suction pressure rises which is the case when the evaporator temperature rises.

The minimum cost of operation occurs when the chilled water supply temperature is maximized, but it must still provide the cooling that is required.

Chilled water pumping costs depend on the temperature difference drop across the load. The higher this difference, the less water that needs to be pumped to transport the required amount of heat. Since the chilled water supply temperature is set by the load, the temperature difference can be maximized only by maximizing the chilled water return temperature.

The optimum setpoint for the chilled water pump is the allowable maximum. Increasing this temperature will lower compressor costs by about 1.5% per F. The heat transfer efficiency of the evaporator is improved by increasing the temperature difference through it. If the temperature difference is 15°F, increasing it by 1°F will lower the chilled water pump cost by about 6%.

With compressor costs of 60% and chilled water pump costs of 15% of the total operating cost, a 1°F increase in the temperature difference will lower the cost of operation by:

$$(1 \times 60)/100 + (6 \times 20)/100 = 1.8\%.$$

An increase in the cooling tower water supply temperature tends to increase the cost of chiller operation but it reduces the amount of work the cooling tower fans need to do. The optimum temperature is the temperature that can satisfy the load at the total minimum cost of all the operating equipment. This is a function of the load and of the weather.

Under some conditions a lowering of the cooling tower water supply temperature can require a large increase in fan operating costs. Data for the application can be based on the measurement of the actual total operating cost or projected costs based on past performance or a combination of these.

Continuous storing and updating of operating cost history as function of load, ambient conditions, and equipment configuration can provide the data required for optimization and also indicate the need for maintenance.

The optimum cooling tower water supply temperature is found by summing the fan, cooling tower water pump and compressor costs. In most cases the major cost is the compressor. There is an increase of about 1.8% in compressor operating costs or 1% of total cooling system cost for each 1⁻F reduction. Pumping costs also increase as the temperature is reduced, but this is only 1/4 or 1/3 of compressor costs unless friction is high from long or undersized piping runs.

If the load is low relative to the size of the cooling towers, the optimum approach corresponds to the safe lowest temperature that the tower can generate. If the load is higher, it will be at some intermediate value.

The optimum return water temperature back to the cooling tower is a function of the cooling tower design. The difference between the supply water temperature and return water temperature is about 15 to 25°F. The setpoint for the cooling tower water pump temperature control becomes the sum of this valve and the supply water temperature.

Water Supply Temperature Optimization

The operating cost of a chiller is reduced by about 2% for each 1°F (0.6°C) reduction in the heat pump temperature difference across this heat pump. In order to minimize the difference, load-following floating control of the chilled water supply can be used. Figure 4-4 shows a system for continuously measuring and maintaining this optimum value for chilled water temperature.

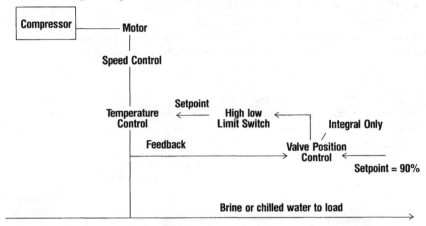

Figure 4-4 Load-following, floating control of chilled-water supply temperature

Energy-efficient refrigeration also depends on equipment sizing and selection. The evaporator heat transfer area should be maximized so that the temperature of the chilled water at the evaporator exit is as close as possible to the average chilled water temperature in the evaporator. At the compressor the refrigerant flow can be made to match the load by motor speed control instead of by throttling temperature control valves. The load-following floating type of control is a more energy-efficient technique.

The chilled water temperature is maximized by selecting the most open chilled water valve and comparing that signal with the 90% setpoint of the valve position controller. If the most open valve is less than 90% open, the setpoint is increased. If the valve opening exceeds 90%, the setpoint is decreased. This allows all users to obtain more cooling by opening their supply valves while the header temperature is maximized.

Lowering the setpoint provides a wider safety margin. Increasing the setpoint maximizes energy conservation at the expense of the safety margin.

Another benefit of load-following optimization involves the reduced valve cycling and lower pumping costs since all chilled water valves are opened and require less pressure drop. Valve cycling is reduced and the valve opening is moved away from the unstable region near the closed position.

In order to keep this type of control system stable, an integral-only controller is used with an integral time that is ten times that of the setpoint controller, usually several minutes. This mode allows the optimization loop to be stable when the valve opening signal is cycling or noisy. High/low limit settings on the setpoint signal guarantee that the setpoint controller will not drive the chilled water temperature to unsafe or improper levels. These limits can block the setpoint controller output so external feedback is used to protect against reset windup.

Water Return Temperature Optimization

The cost of operating the chilled water pumps and the chiller compressor depend on the temperature drop across the evaporator. An increase in this difference decreases the compressor operating costs since the suction pressure rises. Decreasing pumping costs relate to a higher difference with less water needed to be pumped.

This difference will be maximum when the chilled water flow rate across the chilled water users is minimum. (Figure 4-5) The supply water is already controlled. This difference will be maximized when the temperature of the chilled water return is maximum.

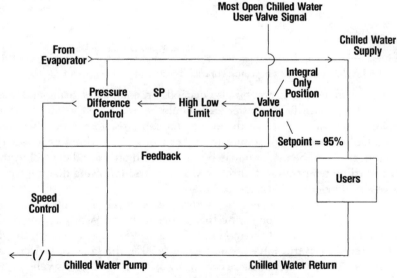

Figure 4-5 Floating control for chilled-water flow rate

If even the most open chilled water valve is not yet fully open, the chilled water supply temperature can be increased by setpoint control, or the temperature rise across the load can be increased by lowering the difference by setpoint control, both methods can be used simultaneously. Increasing the chilled water supply temperature reduces compressor operating cost by about 1.8% for each F of temperature increase and lowering the difference reduces pump operating costs.

The two setpoints determine the sequence of operations and allow the more cost-effective or safer correction to be fully exploited before the less effective one is started. No user valve is allowed to open fully and go out of control as long as the pumps and chillers are sized so as to be capable of meeting the load.

In the chilled water pump control the valve position controller is the cascade master of the pressure difference controller. This guarantees that the pressure difference between the chilled water supply and return is always high enough to supply enough water flow to the load but not so high as to exceed the pressure ratings. The high and low limits are set and the valve position control is free to float the setpoint within these limits to keep the operating costs at a minimum. In order to protect against reset windup when the output of the controller reaches one of these limits, external feedback is provided from the pressure difference output signal to the pump speed controller.

The valve position controller is integral-only, which is tuned to be much more responsive than the valve position controller of the compressor chilled water supply control.

If the chilled water pump station consists of several pumps, only one needs to be variable-speed. Additional pump increments are started when the pump speed controller setpoint is at its maximum. When the load drops, the excess pump increments are stopped on the basis of flow, detected by a low flow switch. In order to eliminate cycling, the excess pump increment is turned off only when the actual total flow corresponds to less than 90% of the capacity of the remaining pumps.

This type of load-following optimization loop floats the total chilled water flow for maximum overall economy. In order to maintain efficient heat transfer and enough turbulence in the evaporator, a small local circulating pump can be used at the evaporator. This pump is started and stopped by low flow switch guaranteeing that the water velocity in the evaporator tubes will never drop below the adjustable limit.

Tower Supply Temperature Optimization

Minimizing the temperature of the cooling tower water is an important part of chiller optimization. Traditional control systems in the past operated with a constant cooling tower temperature of 75°F (23.9°C) or higher.

Each 10°F (5.6°C) reduction in the cooling tower water temperature reduces the operating costs of the compressor by about 15%. If the compressor is operating at 50°F (10°C) condenser water instead of 85°F (29.4°C) for the same load it will use half as much power. Operation at condenser water temperatures of less than 50°F (10°C) is practical during the winter months and savings of 50% are possible.

An optimization control loop can maintain the cooling tower water supply continuously at a minimum temperature which depends on the wet-bulb temperature of the atmospheric air. The cooling tower cannot generate a water temperature that is as low as the ambient wet bulb, but it can near it. As the difference between these temperatures increases, the cost of operating the cooling tower fans drops and the cost of pumping and compressor operation increases. The point where these cross is a minimum point that allows operation at an overall minimum cost. This difference becomes the setpoint of a temperature difference controller which monitors the wet-bulb and cooling tower water supply temperatures. The temperature difference controller output goes to the cooling tower fan speed controller. The cooling tower fans may be centrifugal units or the blade pitch may be variable. Control is maintained by continuous throttling in this case. If the fans are two-speed or single-speed units, the output of temperature difference controller will start and stop the fan units.

Tower Return Temperature Optimization

Optimum control depends on the tower design, and the difference between the supply and return temperatures will be 10 to 25°F. The optimum temperature occurs at the point where the fan cost curve meets the compressor and pump cost curve. The optimized control loop will float the cooling tower water flow rate. A pressure controller guarantees that the pressure difference between the supply and return cooling tower water flows is always high enough to provide flow to the load but not so high as to cause damage. The high and low limits are set on limit switches. A temperature difference controller freely floats the setpoint within these limits, to keep the operating cost at a minimum. In order to protect against reset windup when the output of the temperature difference controller reaches one of these limits, external feedback is provided from the pressure controller output signal to a pump speed controller.

In order to maintain efficient heat transfer and appropriate turbulence in the condenser, a small local circulating pump is used at the condenser. This pump is started and stopped by a low flow switch guaranteeing that the water velocity in the condenser tubes will never drop below the adjustable limit.

Heat Recovery Optimization

When the heat pumped by the chiller is recovered in the form of hot water, a load-following control loop like the control system used for chilled water temperature can be used. If it is possible to operate with 100°F (37.8°C) instead of 120°F (48.9°C) hot water, this will allow the refrigeration load to be met by the chiller at 30% less. This is because the compressor discharge pressure is a function of the hot water temperature in the condenser.

The control loop allows all hot water users to have enough heat while the water temperature is minimized. The most open hot water valve is compared to a 90% setpoint. If the most open valve is less than 90% open, the setpoint of the temperature controller is decreased. If the opening exceeds 90%, the setpoint is increased. This allows all users to obtain more heat by opening their supply valves. Steam heat may be used to supplement the recovered heat to meet the heat load.

A local circulating pump is started when the flow velocity is low. This prevents the formation of deposit in the tubes. This small 10 to 15-hp pump is operated only when the flow is low. The main cooling tower pump is usually larger than 100 hp is stopped when the pump is on.

Optimized Operating Mode

Heat recovery is a function of the outdoor temperature, the unit cost of energy from the alternative heat source, and the percentage of the cooling load that can be used as recovered heat. Depending on the cost of steam and the part of the load needed in the form of hot water, it may be more cost-effective to operate the chiller on cooling tower water and use steam as the heat source when the outside air is below about 65°F (18°C).

When the outdoor temperature is above 75°F (23.9°C), the penalty for operating the condenser at hot water temperatures is not as great and recovered heat should be used. This type of cost-benefit analysis can be continually done with a computerized system.

In warm locations, there may be no alternative heat source and all the heating needs must be met by recovered heat from the heat pump. During cold winter days there may not be enough recovered heat to meet the load. An artificial cooling load is placed on the heat pump. This artificial heat source may be the cooling tower water.

A direct heat exchanger between cooling tower water and chilled water streams may also be used when there is no heat load but when there is a small cooling load during the winter. The chiller can be stopped, and the small cooling load can be met by direct cooling from the cooling tower water that is at a winter temperature. The cooling tower water can be used as a heat source with the compressor on or as a method of direct cooling with the compressor off.

Reconfiguration Optimization

Another approach to optimization is to reconfigure the system in response to changing loads, ambient conditions, and utility costs. During some operating and ambient conditions, the cooling tower water may be cold enough to meet the load directly. If the cooling tower water temperature is below the temperature required, the chiller can be operated in a thermosiphon mode. The refrigerant circulation is then driven by the temperature differential instead of the compressor. In the thermosiphon mode of operation, the chiller capacity drops to about 10% of its rating.

Chiller or heat pump operation is often called mechanical refrigeration. When the cooling takes place directly by the cooling tower water, without the use of chiller or refrigeration units, this is called free cooling. Today, most cooling systems have the capability to operate in several different modes as a function of load, ambient conditions, and utility costs. The switching from one operating mode to another is usually based on economics.

The control system select the mode of operation that will satisfy the load at the lowest cost. The reconfiguration of the system is with valve control. A large complex system can have several dozen modes of operation.

Free Cooling

Free cooling can be direct or indirect. Indirect free cooling maintains a separation between the cooling tower water and the chilled water loops. Table 4-4 shows the valve controls that would be used for a system with four basic modes of operation.

Table 4-4 Cooling Modes

	Mechanical refrigeration	Vapor migration free cooling	Indirect free cooling with heat exchanger	Direct free cooling with filtering
Compressor	0	0	0	0
Cooling-tower pumps	X	X	X	X
Chilled-water pumps	X	X	X	0
Condenser flow	X	X	0	0
Migration Valves	0	X	0	0
Heat Exchanger	0	0	X	0
Evaporator	0	X	0	0
Filter	0	0	0	X
Chiller pump bypass	0	0	0	X

Mode 2 of the table illustrates an indirect free-cooling configuration where the compressor is off. The heat is transferred from the evaporator to the condenser using the natural migration of refrigerant vapors.

Refrigerant migration valves are used to equalize the pressures in the evaporator and condenser. Since the condenser is at a lower temperature, the refrigerant that is vaporized in the evaporator is recondensed there. It is returned to the evaporator by gravity flow. The cooling capacity of a chiller is about 10% of maximum rating in this mode of operation.

The cooling system can be operated in this mode when the load is low and the cooling tower water temperature is about 10°F (5.6°C) below the required chilled water temperature. This usually requires 40 to 45°F (4.4 to 7.2°C) cooling tower and 50 to 60°F (10 to 15.6°C) chilled water temperatures. The high chilled water temperatures are more likely in the winter, when the air is dry and dehumidification is not needed. This mode is implemented with both sets of pumps on and the temperatures are floated as a function of the load and ambient conditions.

A differential temperature controller operates the cooling tower fan based on the wet-bulb and tower water supply temperatures. The setpoint is minimum based on fan costs and compressor and pump cost. This is the floating technique discussed earlier. Another differential temperature controller operates the cooling tower pump. These controls can be overridden when the refrigerant migration valves approach their full opening. This indicates that increased heat transfer is needed at the condenser to meet the load.

The migration valves are throttled by a temperature controller whose setpoint is maximized by a valve position controller which allows all user valves to open until the most-open valve reaches a 90% opening. The chilled water pumping rate is minimized by another valve position controller which provides the setpoint for controlling the chilled water pump. This valve position controller normally keeps the most open user valve at 90% opening, the setpoint for the chilled water pump is not reached under normal conditions. The setpoint is kept at its allowed minimum which results in minimizing the pumping costs. If any of the user valves opens to over 95%, the pumping rate increases to guarantee sufficient coolant to that valve. At this point the system is using the cooling capacity stored in the circulated chilled water. Once this is exhausted, control will be lost. To handle this condition, it is required to switch to a different mode of operation that can provide more cooling.

Plate and Frame Heat Transfer

Another technique for indirect free cooling involves bringing the cooling tower water and the chilled water into heat exchange using a plate and frame type heat transfer unit. This mode allows heat to be transferred from the cooling tower to the chilled water system bypassing refrigeration operation. The heat transfer capacity depends on the size of the heat exchanger plates. Generally, this mode is not restricted to loads of 10% or less like vapor migration.

Indirect free cooling using the plate-type heat exchanger can be operated manually without any controls. Here, the chilled water temperature floats with the load and ambient temperature and the pumps operate at full capacity.

A control system can optimize the operation in this mode. The pumping rate of cooling tower water circulation is set by the highest of three controller outputs. A pressure controller is used to guarantee the minimum pressure differential required for the exchanger. Overrides are implemented if more cooling is required.

Setpoint control is used to prevent the most-open user valve from exceeding a 90% opening. Under normal conditions, the pumping rate of chilled water is kept at a minimum, which maintains the minimum pressure difference across the load. When the controller is unable to keep the most-open user valve at a 90% opening and it rises to 95%, the pumping rate is increased. The added cooling capacity is available at the expense of heating the stored chilled water in the piping system. If this condition occurs for more than several minutes, the system should be switched to a cooling mode that can handle the increased load. This can be mechanical refrigeration or direct free cooling.

Direct Free Cooling

The cooling tower water can be piped directly to the load. This is the most cost-effective mode of cooling, since both the compressor and the chilled water pumps are off. This mode of operation can handle high loads. It is limited only by the size of the towers and their pumps.

Direct free cooling can bring dirty cooling tower water into the system, causing plugging and buildup on the heat transfer surfaces. This can be solved either by full flow filtering (strainer cycling), or by the use of closed-circuit, evaporative cooling towers. In these noncontact or closed-loop cooling towers, the water has no chance to pick up contaminants from the air.

This configuration can operate without automatic controls. Here, the cooling water temperature floats as the load and ambient conditions change. The fan and pumping rates are not optimized. Optimization controls include temperature difference control for the cooling tower fan and pump. Override control is used when a higher pumping rate is needed. Setpoint optimization is used to keep the most-open user valve from exceeding a 90% opening.

Mode Reconfiguration

Some complex cooling systems may include optional storage and heat recovery systems along with alternate types of motor drives. The number of possible modes of operation can get quite high. Reconfiguration may require equipment to be started or stopped, control loops to be reconfigured and pump discharge heads to be modified. As switching occurs, water circulation loops can become much longer. This shifts the operating point of pumps and can lower their efficiency.

Control loop requirements for the four basic modes of operation are shown in Table 4-5.

If the control loops are automatically reconfigured, their tuning constants are changed as their outputs are switched to different manipulated variables, since the time constants of the loop are also changing.

The switching from one mode to another should be done in the off state, if an interruption of a few minutes can be tolerated. The pumps serving the load can be left on, using the coolant storage capacity of the water distribution piping. The three-way valves used will never block the pump discharge but will gradually change the water flow.

If the mode changes are frequent or if the coolant capacity in the piping is too small to meet the load requirements for a few minutes, the system must be switched while running. This requires a higher level of automation, since the starting of some equipment requires more safety interlocks.

Automatic mode reconfiguration is a powerful technique for optimization, since it makes the system flexible. It is not limited to cooling systems but can be used for any operation where the system must adapt to changing conditions.

Table 4-5 Control Loop Requirements

Control	Mechanical Refrigeration	Vapor Migration	Indirect	Direct
Temperature Difference	Fan	Fan	Fan	Fan
Temperature Difference	Tower Pump	Tower Pump	Tower Pump	Tower Pump
Pressure Difference	Tower Pump	Tower Pump	Tower Pump	Tower Pump
Valve Position		Pressure Difference Setpoint		
Temperature	Compressor	Migration Valves	Tower Pump	Tower Pump
Valve Position	Temperature Setpoint	Temperature Setpoint	Temperature Setpoint	Temperature Setpoint
Valve Position	Pressure Difference Setpoint	Pressure Difference Setpoint	Pressure Difference Setpoint	
Pressure Difference	Chiller Pump	Chiller Pump	Chiller Pump	

Optimized Storage

If storage tanks are used, the daily brine or chilled water needs can be generated at night, when it is the least expensive. The ambient temperatures are lower and off-peak electricity is less expensive in some areas. When demand is low, operating costs may be lowered by operating the chillers part of the time at peak efficiency rather than continuously at partial load. Efficiency is reduced at low loads due to losses from friction drops.

Chiller cycling allows the thermal capacity of the chilled water distribution system to absorb the load while the chiller is off. A pipe distribution network with a volume of 10,000 gallons (37,800 l) represents a thermal capacity of about 100,000 BTUs for each F of temperature rise.

If the chilled water temperature is allowed to rise 5°F (2.8°C) before the chiller is restarted, this represents about 40 tons (140 KW) of thermal capacity. If the load is 20 tons (70 kW), the chiller can be off for two hours. If the load is 80 tons (351 kW), the chiller can be off for only 2 minutes. If the chiller needs a longer time off, the tankage can be increased with more water volume. Another chiller can be started or the load can be distributed among several chillers.

Load Allocation

The continuous measurement of the actual efficiency of each chiller allows the loads to be met using the most efficient combination of machines for the load. The cost per ton of cooling can be calculated from measurements and used to establish the most efficient combination of units to meet the loads.

In simple load allocation, only the starting and stopping of the chillers is optimized. When the load is increasing, the most efficient idle chiller is started. When the load is dropping, the least efficient one is stopped. In systems with more automation, the load distribution between chillers is also optimized. A computer calculates the real-time efficiency of each chiller. A calculation of the incremental cost for the next load change for each chiller is made. If the load increases, the incremental increase is sent to the setpoint of the most cost-effective chiller. If the load decreases, the incremental decrease is sent to the least cost-effective chiller (Figure 4-6). The software does continuous load balancing and makes predictions of costs and efficiencies.

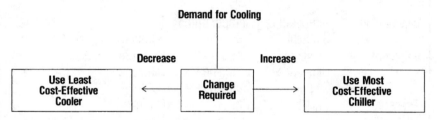

Figure 4-6 Computer-based load allocation

The most efficient chiller will tend to reach its maximum loading or enter a region of decreasing efficiency and will no longer be the most efficient. When the loading limit is reached on a chiller, the computer selects another as the most efficient unit.

The least efficient chiller accepts the decreasing load signals until the minimum limit is reached. Its load will not be increased until all other chillers are at their maximum load.

Some chillers have high efficiency at normal load while being less efficient than others at low loads. These units are not allowed to shut down, but are given a greater share of the load.

When evaporators are operated in parallel from the same compressor, the load can be followed by varying either the evaporator temperature or the refrigerant level in the evaporators. One control scheme is to keep the evaporator wetted heat transfer areas constant and let the load variations result in modifications in the evaporator temperature. The most open temperature control valve will not open to more than 90% if the compressor speed is increased when the valve opening reaches 90%. This results in a reduction in compressor suction pressure which lowers the evaporator temperature and increases the rate of heat transfer. No user is ever allowed to run out of coolant, and the compressor operating costs are kept at a minimum.

This type of load-following optimization, where the compressor speed varies with the requirements of the most heavily loaded user, is very sensitive to dynamic upsets. The compressor controller must be tuned to allow slow changes in the compressor speed to avoid upsetting other users which are not selected for control.

Another technique is to do away with any vapor-side control valves. Heat transfer is modulated by adjusting the tube surface area exposed to boiling. The user with the highest load will have the highest refrigerant level. The most flooded evaporator is used for control to increase the speed of the compressor when the level exceeds 90% and lower it when the level is below 90%.

This type of load following wastes that part of the heat transfer surfaces in the evaporators that are not fully utilized. Another problem involves foaming that can mask the level of heat transfer area. This can keep the tubes wet even when the level has dropped. Tuning can also be difficult and it may be better to keep the refrigerant level constant and control the evaporating pressure.

Feedforward Control

Feedforward control can increase the responsiveness of the control system and also provide more precise temperature control. This precision is needed when the goal of optimization is to maximize chilling without freezing. The load is found by multiplying the flow rate of the cooled fluid by its drop in temperature. Based on the load and the desired temperature, the refrigerant temperature is found.

Retrofit Optimization

In new installations, variable-speed pumps can provide the thermal capacity required for chiller cycling. Chillers and cooling towers can be located at the same elevation and near each other so that pumping costs will be minimized. In existing installations, there are inherent limitations in what can be done.

In optimizing existing chillers, the constraints include low evaporator temperature, economizer flooding, surge, and piping/valving limitations.

Surge occurs at low loads when not enough refrigerant is circulated. A surge condition can cause violent vibrations and eventual damage. Most older chillers do not have automatic surge controls and may have only vibration sensors for shutdown. If the chillers operate at low loads, an anti-surge control loop is needed.

Surge protection works against efficiency. To bring the machine out of surge, the refrigerant flow must be artificially increased if there is no real load on the machine. This increase in flow is artificial and wasteful usually involving water bypasses. A more economical method is to cycle a large chiller or to operate a small unit to meet low load conditions.

Low temperatures can occur in the evaporator when an older chiller is optimized. A chiller that has been designed for operation at 75°F (23.9°C) condenser water is run at 45°F or 50°F (7.2°C or 10°C) in the winter.

This is the opposite of surge, since it occurs when refrigerant is being vaporized at an excessive rate. The vaporization occurs because the chiller is pumping more than its design load because of the low compressor discharge pressure. At this point, the only way to increase heat flow is to increase the temperature differential across the evaporator tubes. This lowers the refrigerant temperature in the evaporator until it reaches the freezing point and shuts down.

Increasing the evaporator heat transfer area involves major equipment modifications. But, preventing the refrigerant temperature in the evaporator from dropping below the freezing point can be done by not allowing the cooling tower water to cool the condenser to its own temperature. This can be accomplished with a temperature control loop that limits the chiller from using all of the available cold water from the cooling tower. The loop will throttle the water flow rate and cause its temperature to increase.

Economizer Control

In most chillers the economizer control valves are sized such that the refrigerant vapor pressure in the condenser is constant and relates to a condenser water temperature of 75°F or 85°F (23.9°C or 29.4°C). When these units are operated with 45°F or 50°F (7.2°C or 10°C) condenser water, the vapor pressure is much lower.

This results in higher circulation rates and the control valves may be unable to provide the required flow rates. Flooding of the economizer can occur with these higher flows. Larger valves may be needed with proportional and integral control modes. Proportional controllers cannot maintain the setpoint as the load changes. The addition of the integral mode eliminates this offset. Otherwise the compressor can be damaged by the liquid refrigerant that overflows into it from the flooded evaporator.

Water Distribution

In the ideal water distribution system, individual users are served by two-way valves and the optimum supply temperature is found by keeping the most open valve at 90% opening. In existing installations, the number of valves may be high and distributed over a large area. If the loops are not properly tuned, the valves may cycle from full open to full closed.

In this case, a few representative user valves that are not cycling can be used for optimization. It may be possible to group the user valves into high and low priority categories. These representative or high priority valves are treated with load-following, floating control. The low-priority valves are grouped together, and their demand is detected by the measurement of total flow. If the low-priority valves are all two-way, and if the supply temperature is constant, the flow will vary directly with the load. If valve-position-based optimization is used, it would attempt to maximize valve opening, which would maximize flow. This same goal can be achieved by raising the supply temperature, which will increase the total flow. Critical users are protected by valve position control, which overrides the flow control signal whenever a high-priority valve reaches 90% opening.

In many air conditioning applications three-way control valves are used. This increases pump costs and lowers return water temperature. A control system can be used for optimization, where the total flow is relatively constant and the return water temperature reflects the variations in cooling load. The setpoint will be a function of the difference between the supply and return temperatures (the temperature rise) and the desired space temperature setpoints.

If the load drops, the space temperature will begin to fall, and the thermostats will divert more coolant into the return line. This will reduce the return temperature. In a conventional system, this would become the new steady state. With an optimizing control, this temperature is adjusted up to save cooling costs.

As the supply temperature increases, the space thermostats will divert less water until they stop at the same percentage of water diverted as at full load. This technique reduces the temperature differential across the chiller, which can lower the operating cost by about 15% for each 10°F (5.5°C).

Valve-position override is used when a high-priority valve is more than 90% open. This overrides the setpoint and lowers it until all high-priority valves are less than 90% open.

The proper control and optimization of chilled water systems also depends on the piping layout. The configuration should provide chilled water storage to handle emergencies or power failures.

Chapter 5

Environmental Controls

Remote controls and automation techniques can be applied for providing a more comfortable environment. The most widespread form of environmental control is the control of temperature. The thermostat is an automatic device for the control of heating and air conditioning equipment. A thermostat can be a simple automated switch or it can be a more sophisticated electronic device with special features. Figure 5-1 shows an electronic thermostat with an indicating light emitting diode. This chapter discusses thermostats and shows several ways for automating temperature control.

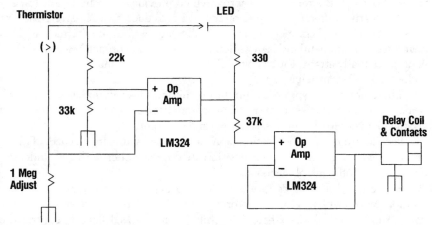

Figure 5-1 Electronic Thermostat

Thermostatic Control

If the space conditions are to be accurately maintained, such as in clean rooms where the temperature must be controlled within +/-1°F (0.55°C) and the humidity must be kept within +/-3% RH, simple types of thermostatic control are unacceptable. In these cases it is necessary to use a resistance temperature detector or semiconductor temperature sensor with a proportional-plus-integral thermostat controller which will eliminate the offset error. This can best be accomplished using microprocessor-based shared controllers that communicate with sensors over data highways. Multistage thermostats are used to operate two or more final control elements in sequence.

A wet-bulb thermostat is used for humidity control. A wick, or other means for keeping the bulb wet, and rapid air motion are needed to assure a true wet-bulb measurement. A dew point thermostat is a separate device designed for control from the dew point temperatures.

Two-position Control

The movement produced by a bimetallic temperature sensor or by the expansion of a humidity-sensitive element can be used to provide a two-position controller. Electric thermostats use various differential gaps for two-position control. One type of thermostat design has the contacts directly on the bimetallic element. The setpoint is adjusted by moving one contact or repositioning the bimetal element. The gap in this device is theoretically zero, but in practice there is a small gap. This contact scheme is known as the zero gap design.

Another design variation for two-position thermostats uses what is called a positive differential gap. This is often better in two-position control since it prevents excessive cycling.

One mechanism uses a toggle action to produce a positive gap. The gap is adjusted through the tension of a spring which provides some hysteresis force.

A negative differential gap is often used to reduce an excessive period of oscillation. The load contacts turn on a small heater within the thermostat. This heat affects the bimetal element which opens the load contact before the temperature actually attains the same point. The load contact make point is normal and is also called anticipation.

There are many types of two-position thermostat or humidostat-type controllers for both domestic and industrial applications. These involve many arrangements of load contacts, setpoint, and differential gap.

For floating controller action a neutral zone is usually needed. This is obtained with a second contact position in the controller with two independently set on-off control mechanisms.

Single-speed floating control is achieved using the same equipment as two-position control with an electric-motor- operated valve. In two-position control the valve stroke is less than 120-seconds while in single-speed floating control the valve stroke is usually 120-seconds or more. An electrical interrupter is sometimes used with the motor to decrease the speed of opening and closing the valve.

Dual Thermostats

Thermostat design variations include the dual thermostat with two setpoints. One setpoint corresponds to the minimum temperature while the other corresponds to the maximum allowable space temperature. These thermostats can use their dual outputs to operate a variable air volume (VAV) damper to cool the space and a reheat coil for heating. The dual outputs can also be used to throttle either the hot or the cold water to condition the air being supplied to the room.

Setback Units

Night-day or setback thermostats can operate at different normal temperature values for day and night. They have both a day and night setting dial and change from day to night operation. In some electric thermostats dedicated clocks and switches are built into each thermostat.

A pneumatic day-night thermostat uses a two-pressure air supply system. The two pressures usually are 13 and 17 PSIG (89.6 and 117 kPa) or 15 and 20 PSIG (103.35 and 137.8 kPa). Changing the pressure from one value to the other actuates switching devices in the thermostat and switches them from day to night. The supply air mains are often divided into two or more circuits so that the switching can be accomplished in various areas at different times.

Heating-Cooling Thermostats

Heating-cooling or summer-winter thermostats can be reversed with the setpoints changed by means by indexing. They can be used to actuate devices, such as valves or dampers, that regulate a heating source at one time and a cooling source at another. Manual indexing may be done in groups by a switch, or automatically by a thermostat that senses the temperature of water supply, outdoor temperature or other variable.

In heating-cooling units there are often two bimetallic elements. One is direct acting for the heating mode and the other is reverse acting for cooling. The mode switching can be done automatically in response to a change in the air supply pressure.

Limited Control

A limited control range thermostat will limit the room temperature in the heating season to a maximum of 75°F (24°C), and in the cooling season to a minimum of also 75°F (24°C), even if the occupant of the room has set the thermostat beyond these limits. This is done internally without a physical stop on the setting.

Zero Energy Band Control

Zero energy band (ZEB) thermostats can provide heating when the zone temperature is below 68°F (20°C) and will provide cooling when it is above a higher temperature such as 78°F (25.6°C). These are adjustable settings. In between the setting neither heating or cooling takes place.

One way to achieve this is to use a dual thermostat. If a single setpoint thermostat is used instead of a dual unit, energy would be wasted, because as soon as the heating is stopped, cooling is initiated. This is wasteful, since no energy needs to be added or removed from a space which is already comfortable (within the ZEB band). This approach can reduce yearly operating costs by about 33%. This type of control can also allow buildings to become self-heating by transferring interior heat to the perimeter.

Slave and Master Thermostats

A slave or submaster thermostat has its setpoint raised or lowered over a predetermined range, in accordance with a master control. The master control may be a thermostat, manual switch, pressure controller, or other device.

A master thermostat measuring outdoor air temperature could be used to adjust a submaster thermostat controlling the water temperature in a heating system. Master-submaster schemes are also known as single-cascade action. When this action is accomplished by a single thermostat with more than one measuring element, it is called compensated control.

Smart Thermostats

Smart thermostats are often microprocessor-based units with an RTD-type or solid-state sensor. It usually has its own dedicated memory and may also have a communication link over a shared data bus to a central computer. These units can minimize building operating costs by combining time-of-day controls with intelligent comfort gaps and maximized self-heating.

Many devices were developed to serve the heating, ventilating, and air conditioning (HVAC) industry when energy costs were low. Performance considerations were secondary compared to the cost of the device. When energy costs went up, higher quality units were developed, but conventional HVAC devices still remain in use in existing systems and in new installations where the installed cost is of primary importance.

Pneumatic Units

A thermostat or humidostat is a type of simplified controller with a proportional control mode. The pressure of the output signal from a pneumatic type of stat is nearly a straight-line function according to the following:

$$S = K_C (T - T_0) + S_0$$

where

S	=	Output signal
K_C	=	Proportional sensitivity, fixed or adjustable
T	=	Measurement temperature (or relative humidity)
T_0	=	Normal value of measurement corresponding to the center of the throttling range
S_0	=	Normal value of the output signal, corresponding to the center of the throttling range of the control valve (or damper).

For a spring range of 3 to 15 PSIG (0.2 to 1.0 bar), S_0 is 9 PSIG (0.6 bar) and for a 9 to 13 PSIG (0.6 to 0.9 bar) spring range, S_0 is 11 PSIG (0.75 bar). Table 5-1 shows the gain and throttling characteristics of a direct acting thermostat.

Table 5-1 Gain and Throttling Range of Pneumatic Thermostats

Range Spring Type	PB (%)	Gain	Proportional Sensitivity PSIG/F	Throttling Range F	Normal Output PSIG	Spring Range PSIG
A	3.2	31	2.5	1.6	5.0	3-7
B	5.0	20	2.5	2.0	7.5	5-10
C	5.0	20	2.5	2.0	10.5	8-13
D	3.2	31	2.5	1.6	6.0	4-8
E	3.2	31	2.5	1.6	11.0	9-13
F	3.2	31	2.5	1.6	9.0	7-11
G	9.6	10	2.5	4.8	9.0	3-15

Setpoints

It is common to call T_0 the setpoint of that thermostat. Strictly speaking, thermostats do not have setpoints since they are simple proportional devices. Integral action is needed in order for a controller to be able to return the measured variable to a setpoint after a load change. T_0 does not represent a setpoint, it only identifies the temperature that will cause an action to take place. This can be called a normal or nominal condition, because relative to this point the thermostat may both increase and decrease the cooling air flowrate as the space temperature changes. Thermostats do not have setpoints, but they have throttling ranges. If the throttling range is narrow enough, this gives the appearance that the controller is keeping the variable near the setpoint, but the narrow range allows the variable to drift within limits.

Throttling Range

The control response of a thermostat can be described by the slope of its operating line. This slope is characterized in different ways depending on the different industry and manufacturer involved. In process control the terms gain and proportional band are used while in HVAC the terms proportional sensitivity or throttling range are more common. All of these terms describe the same slope.

The throttling range is the range in which the space conditions are allowed to drift as the final control element is modulated from fully closed to fully open.

Proportional sensitivity is the amount of change in the output signal that results from a change of one unit in the measurement of temperature or humidity.

Gain is the ratio between the changes in control output and measurement input. For example, suppose a 100% change in control element output results in

2°F change in measurement. This represents 5% of the thermostat span of 50 to 90°F and the gain of this thermostat is 20.

The proportional band is the percent change divided by the gain:

PB = 100/G = 100/20 = 5%

Thermostat Action

A thermostat is said to be direct acting (D/A) if its output signal increases as the measurement rises. A reverse-acting (R/A) thermostat changes the slope of the operating line from upward to downward.

The operating line will vary with the spring range of the final control element. Proportional sensitivity (gain) and throttling range change with the different type of springs. Typical spring ranges and applications are listed below:

Fail Closed HVAC quality dampers
A-3 to 7 PSIG (0.2 to 0.5 bar)
B-5 to 10 PSIG (0.35 to 0.7 bar)
C-8 to 13 PSIG (0.55 to 0.9 bar)

HVAC quality valves
D-Fail Open, 4 to 8 PSIG (0.9 to 0.55 bar)
E-Fail Closed, 9 to 13 PSIG (0.6 to 0.9 bar)
F-3-way, 7 to 11 PSIG (0.5 to 0.76 bar)

Failed Closed Speed or Blade Pitch Positioners
G-3 to 15 PSIG (0.2 to 1.0 bar)

The wider the throttling range, the larger the control error which is an offset or drift away from the normal measurement value. A wider throttling range does provide a more stable system with less cycling of the final control element. A tradeoff exists between lower stability or higher offset error. With their spring ranges of 4 or 5 PSIG (0.3 or 0.35 bar), most conventional HVAC controls lean towards lower stability as the more prevalent of the two options. In process control, just the opposite is true. This choice stems from the need to maximize stability. Offset error is less of a concern since it can be eliminated by using the integral control mode.

Sensitivity

The simplest thermostats are made with fixed sensitivities. The control flexibility is always better when the proportional sensitivity is adjustable. A typical 5:1 adjustment range can be found in standard thermostats. A lower proportional sensitivity improves stability, but it also increases the offset error due to a wider throttling range.

Thermostat Design

The basic pneumatic thermostat with fixed-proportional control uses a flapper arrangement. The flapper moves in front of a nozzle to affect a low volume output signal to dampers or control valves. The temperature detection is done with a bimetallic element which changes the flapper position relative to the manually adjusted normal value. Air at about 20 PSIG is supplied to the open nozzle and the backpressure is inversely proportional to the distance between nozzle and flapper.

The proportional sensitivity K_C is the change in output pressure per F measured. This is a fixed value when the ratio of A/B is fixed, while it is adjustable if the ratio of A/B is adjustable.

Due to the small restriction and nozzle openings used, the resulting output airflow is rather small at around 1.0 SCFH (0.018 m^{3}hour). This type of design is also referred to as nonrelay or low volume. The 1.0 SCFH air capacity is usually not enough to operate final control elements. In order to increase the air capacity of a thermostat, a booster or repeater relay is added. The relay type or high volume humidostats or thermostats can provide an output airflow of 10 to 30 SCFH (0.3 to 0.84 m^3/hour). In the relay-type, the nozzle backpressure does not operate a final control element directly, but it is carried to a bellows chamber and acts against the area of the bellows.

Humidostat Design

These units are similar to thermostats, except that the bimetallic measuring element is replaced by a humidity-sensitive element. A typical humidity-sensitive element is cellulose acetate butyrate. This substance expands and contracts with changes in relative humidity, and the resulting movement can be used to operate the flapper of a pneumatic humidostat or to open and close the contacts of electric humidostats.

The choice of features is similar to those of thermostats. Humidostats can be direct or reverse acting and they can be high or low volume (relay type or nonrelay type). The normal value can be set manually or by remotely adjustable pneumatic signal.

The proportional sensitivity determined by the A/B ratio can be fixed or adjustable. A feedback bellows can be used to minimize variations in supply pressure or temperature and to increase the range of adjustability of the proportional sensitivity setting.

Humidostats are often installed in air ducts to control upper and lower limits of the relative humidity of the air. These devices are usually set for either 85% RH or for 35% RH.

Thermostat Accuracy

In conventional HVAC pneumatic thermostats and humidostats, the total error is the sum of the sensor error and offset error. Even with individual calibration, the sensor error usually cannot be reduced to less than +/– 1°F (0.55°C) for thermostats or +/– 5% RH for humidostats. Added to the sensor error is the offset error, which increases as the throttling range increases and which can be as high as +/– 5°F (2.8°C) for thermostats and +/– 20% RH for humidostats. The total errors can approach +/– F (3.3°C) and +/– 25% RH.

In conventional HVAC electrical thermostats or humidostats, the total error is made up of three components. There is the element error, which is the same as the sensor error in pneumatic units. Then, there is the mechanical differential set in the thermostat switch, which controls the final control element by turning it on and off. This differential or bandwidth can vary for thermostats from 0.5 to 10°F (0.3 to 5.6°C) and for humidostats from 2 to 50% RH.

The third error component is due to the time lag from turning on the heat or humidity source and returning the space to within the control differential. The thermal conditions will continue to deviate further while the system is on. The total error from these three factors can be as high as for HVAC pneumatics. However, if the final control elements are large enough and the control differential is set small, most electrical units can outperform pneumatic devices.

Temperature Sensors

There are several approaches to electrically monitoring temperature. Commonly used temperature sensors in electronic circuits include thermocouples, diodes and thermistors. Thermocouples are one of the simplest electrical temperature sensors. It consists of a junction of two wires made of different metals. When the junction is heated, a voltage proportional to the applied temperature is developed across the junction. This is known as the Seebeck effect. The developed voltage can then be measured and used to control different electrical circuits, as desired.

A standard silicon diode can also be used as a temperature sensor if a small forward bias is applied to the diode. The voltage drop across the diode will respond to changes in temperature at a rate of about 1.25 mV per degree F.

A diode temperature sensor circuit is shown in Figure 5-2. This circuit is a variation on the Wheatstone bridge which is used in measurement circuits of many different types.

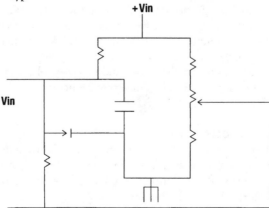

Figure 5-2 A diode used as a temperature sensor

The supply voltage for this circuit should be between + 1 and + 1.5 V. Almost any silicon diode can be used in this type of application, including the inexpensive 1N914 diode.

Thermistors

Another popular type of temperature sensor is the thermistor, or thermal resistor. All resistance elements are temperature sensitive, but a thermistor is designed to emphasize the effects of temperature.

There are two basic types of thermistors:

1. NTC thermistors are the more common, with a negative temperature coefficient. This means that the resistance decreases as temperature increases.
2. PTC thermistors have a positive temperature coefficient and operate in the opposite way. The resistance increases as the temperature increases.

Thermistors have their limitations, in most cases, their temperature-resistance curves are not linear. This makes it difficult to design a thermistor-based circuit that is linear over a wide range although many electronic thermometer circuits have been designed with thermistors. NTC thermistors can exhibit negative resistance characteristics under some conditions. This can result in thermal runaway and possible damage to the thermistor and other circuit elements.

Ohm's law specifies the relationships between resistance, current, and voltage:

$$R \quad = \quad E/I$$
$$E \quad = \quad RI$$
$$I \quad = \quad E/R$$

where

$$R \quad = \quad \text{resistance}$$
$$I \quad = \quad \text{current}$$
$$E \quad = \quad \text{voltage.}$$

Suppose an NTC thermistor is placed in series with a fixed resistor. Since it is very small compared to the thermistor, we can ignore the thermal drift of the fixed resistance. Let the applied voltage be 10 V and the fixed resistance be 1000 ohms. The resistance of the thermistor is dependent on the temperature and the higher the temperature, the lower the resistance, since this is an NTC device.

At some temperature, the thermistor will have a resistance of 10,000 ohms. The total resistance of the circuit is 11,000 ohms. The current flowing through the circuit will be equal to

$$I \quad = \quad E/R = 10/11{,}000 \approx 0.00091 \text{ A} = 0.91 \text{ mA}$$

The voltage drop across the fixed resistance is approximately equal to

$$E \quad = \quad RI = 1000 \times 0.00091 \approx 0.91 \text{ V}$$

The voltage drop across the thermistor will be about

$$E \quad = \quad RI = 10{,}000 \times 0.00091 \approx 9.1 \text{ V}$$

Because of the negative temperature coefficient, the resistance of the thermistor will drop as the temperature increases. At some higher temperature, the thermistor's resistance will be 1000 ohms making the total resistance in the circuit equal to 2000 ohms. Now the current flowing through the circuit is about

I = 10/2000 = 0.005 A = 5 mA

Note that the current has increased with the temperature and the voltage drop across the two resistance is now equal because of their equal values:

E = 1000 x 0.005 = 5 V

As the temperature increases, the current flow and the voltage drop across the thermistor also increases. The voltage across a resistance element is dissipated as heat and the higher the voltage drop, the greater the heat produced.

If the ambient temperature increases, the resistance of the thermistor drops and the voltage dropped across it increases. The voltage drop across the thermistor generates heat which increases the temperature sensed by the component. This reduces its resistance, which increases the voltage drop and increases the generated heat. A thermal runaway condition exists unless the current is kept low enough to protect the components.

It may result in a slight loss in accuracy. Real harm is done when large currents flow through an NTC thermistor. The sensed temperature then becomes meaningless and may damage the control circuitry.

PTC thermistors are not subject to thermal runaway. When the temperature increases, the resistance follows reducing the voltage drop and decreasing the heat generated. Because of this, PTC thermistors can be used as temperature regulators. When power is applied, the thermistor will heat itself until a stable level is reached. An almost constant temperature will be maintained as changes in the ambient temperature or thermal load occur.

If the thermistor is moved from one temperature extreme to another, it cannot respond instantly. The mass of the thermistor's body must heat up or cool down to match the new ambient temperature. This takes a finite amount of time.

The larger the thermistor's body, the greater the time required for it to respond to a change in ambient temperature. For a small disc or bead thermistor, the response time will typically be a few seconds. Larger thermistors will have a longer response time. Some of the larger thermistors have a body that is 1-inch in diameter. These devices can require several minutes to respond to a change in the ambient temperature.

Temperature Probes

In most cases, the temperature sensing element will be housed in some type of protective probe assembly. This is especially true if high humidity is present. A conductive liquid or vapor can cause leakage between the connecting leads, resulting in false readings. Corrosion can affect the leads or the sensor body.

Except for hermetically sealed glass body thermistors, liquid can creep inside the component itself, seeping through minute openings between the leads and the epoxy coating. This can result in misreadings and eventual corrosion of the sensing element.

Also, an unshielded temperature sensor can be cooled by wind. This is desirable if you want to monitor the wind chill. If you are interested in the actual temperature, it can be problem.

The solution to these problems is to house the sensor in some type of enclosure. An isolated temperature sensor can be enclosed in a probe housing. The probe may be a tube of glass or stainless steel welded shut at one end. The connecting leads are soldered to the sensor and brought out through the back of the probe. A threaded probe can be screwed into a socket or holder.

Electronic Thermostats

One type of electronic thermostat circuit for controlling an electric heating element uses a triac, SCR and IC temperature sensor like the LM3911. In this type of electronic thermostat the temperature sensor controls the SCR which, in turn, controls the triac.

This type of thermostat circuit can directly control an electrical heating element of 50 A. The triac must be rated to handle the load current. Some circuits use an electronic fuse which opens the load when the current rating is exceeded.

A comparator circuit can also be used as an electronic thermostat. A thermistor or other temperature sensor shunts the current across the comparator input. One connection method grounds one end of the sensor with the sensitivity control resistance across the comparator input. Another method reverses these devices in the electronic thermostat circuit.

If the monitored temperature goes above a preset level determined by the values of the shunt resistors, the comparator's output goes high. The output of the comparator is fed through a buffer which is often a simple noninverting voltage follower.

A relay is connected to the buffers output. When the comparator's output is high, the relay is on. If the monitored temperature is below the reference point, the comparator's output goes low and the relay is switched off.

The thermostat can control a heating device using the normally open contacts of the relay. A cooling device can be controlled by using the relays normally closed contacts instead of the normally open contacts. A diode is often used to clamp the input to the relay. This clamps any voltage transients due to the inductive storage of the coil.

Most heating and cooling devices use fairly heavy currents and the output might not be able to control a large enough relay directly. An electronic relay could be used or two relay stages could be connected to deliver larger currents.

Comparators are also used for simple over/under temperature alert that indicate when a monitored temperature is within a preset range. In this type of circuit two comparators are connected in series so they can also indicate if an out-of-range temperature is too high or too low.

LEDS (Light Emitting Diodes) can be used to indicate the monitored temperatures relationship to the preset range. The comparator signals can also be buffered and sent to a computer for HVAC control.

The control output or LED will indicate:
  • Temperature too high,

- Temperature in range or
- Temperature too low.

The reference range depends on voltage divider resistors, along with the characteristics of the temperature sensor used.

Boiler Control

In a boiler based heater, the system's water temperature is controlled by a device called an aquastat. Greater fuel efficiency can be achieved using a lower water temperature setting in summer than in winter. Typically, these values are about 180°F in winter and 160°F in summer. The water temperature varies inversely with the outside temperature. Some controllers monitor the outside temperature and automatically adjust the water temperature accordingly. The circulator control is usually set to about 125°F to 135°F. Rather than using a seasonal approximation, the water heater can respond directly to changes in the weather, maximizing fuel efficiency.

Two temperature sensors are used, one for air and the other for water. A solid state sensor such as the AD590 or similar device may be used. A switch allows the user to select between automatic and manual (override) modes. The output relays switch contacts are wired in series with the boiler's aquastat. The lower and upper temperature limits may be set with potentiometers. This is performed with both sensors at the same temperature using an accurate reference thermometer. The two sensors and the thermometer need to be in a dry, even-temperature location that is protected from significant breezes.

Temperature Equalization

A thermostat is a closed-loop type of control circuit since there is a continuous circular path throughout the system. The thermostat monitors the room temperature and controls the furnace and the heat from the furnace changes the room temperature, which affects the thermostat. The input is the room temperature and the output is the heater. These are interrelated, affecting each other directly. Because of this interrelation between the input and the output, the closed-loop system can exhibit some instability or oscillation.

Suppose the thermostat is located some distance from the heater duct and there is only limited air circulation in the room. The temperature in the room drops, so the thermostat directs the furnace to put out more heat. Because of the distance from the heater duct to the thermostat, it will be some time before the temperature at the thermostat is high enough for it to shut off the furnace. Part of the room becomes too hot.

When the furnace is turned off, the room begins to cool, especially near heat leaks, such as doors and windows. Assume that these leaks are also a significant distance from the thermostat. It will take some time before the thermostat can sense the drop in temperature. By this time, part of the room might become too cold.

The room's overall temperature oscillates from too hot too cold, and back again, without reaching an optimum value. There is a loss of the proper heating benefits of the system as well as a sizable waste of fuel or electrical power due to these unnecessary oscillations.

One course of action would be to position the thermostat closer to the heating register and the heat leaks. Because the thermostat can only monitor its own location in the room, hot or cold spots can still develop in other areas. A better reaction is to increase the motion of the air. The air motion will tend to stabilize the temperature. Warmer air can be moved into colder areas and colder air into warmer areas. A temperature sensing fan control can be used to control a fan in response to changes in ambient temperature.

In this type of control a special purpose IC such as an LM3911 may be used to sense the temperature. A temperature equalizer, such as this helps to get the maximum benefit out of the heat/cooling system. The LM3911 temperature sensor control IC may be used to operate a relay driver made of 2N2222 transistors or similar devices. The relay, which can also be a solid state device, must be able to safely carry the current drawn by the fan. The temperature that switches on the fan may be determined by a potentiometer that controls the sensor input.

A continuously running fan will also stimulate air movement, but at the cost of wasted power and increased ambient noise. It could also produce drafts in some areas. The controller allows the fan to only be turned on when it is needed.

Heater Humidifiers

In the winter months heaters are generally in heavy use, increasing fuel usage and driving the indoor humidity down. This decreased humidity has several effects. Air is a poor conductor, which means a greater buildup of static electricity.

The human body is humidity sensitive. The most healthy environment has humidity levels in the 30 to 50% range. In winter the house is sealed off, there is limited air exchange with the outdoors, and heating units dry up the humidity in the air. The indoor humidity in winter can drop to 10 to 20% which is extremely dry. This can irritate sensitive membranes leading to sore throats and other ailments. Also, the body's humidity sensitivity can make dry air in the 50 to 70°F range seem even colder than normal. Raising the humidity to 30 to 50% allows a lower thermostat setting for the same degree of comfort. Too much humidity can be as bad as too little. Humidity levels greater than 60% can significantly increase discomfort, and make cold temperature seem even colder. Also germs tend to thrive in moist air.

A humidifier is a circuit that will add humidity to heated air. A temperature sensor such as a LM3911 IC is placed in the air path of the heaters vent. When the temperature exceeds a point set by a potentiometer, a solenoid is activated, opening a spray nozzle. The water is sprayed directly into the heating vent, where it is immediately vaporized, adding to the air humidity.

A circuit that adds humidity to dry heated air might use the LM3911 temperature sensor control IC along with a driver transistor like the 2N3906 to operate a triac or similar device. The triac drives a solenoid which controls the water supplied to the spray nozzle. The spray nozzle sprays a relatively fine mist of water for easy and quick vaporization.

Air Conditioner Humidity Controls

Excess humidity can be a problem in summer. The excess humidity makes hot temperatures seem even hotter. An air conditioner not only cools the air, it also drains off excess humidity. In relatively low humidity, 80°F can seem cooler than 77°F in high humidity. Setting the thermostat too low is a waste of energy and may result in too much cooling at certain times.

The body responds to a combination of temperature and humidity while a thermostat responds to temperature alone. It does not sense the humidity at all. A humidity controller will cycle the air conditioner's compressor to minimize humidity. This keeps the room cool while the air conditioner will actually have to run less.

Most air conditioners use 24-V dc control circuitry. The connections to the air conditioner's thermostat may be indicated by small circles containing one of the following letters: G, R, W, and Y. These letters indicate a standard color coding as shown in Table 5-2. This table also indicates alternate labeling schemes used in commercial air conditioner thermostats. The green wire should run to the fan and the yellow wire should run to the compressor.

Table 5-2 Thermostat Connections

Schematic Label	Cable color	Typical Thermostat Labels				
G	Green	G	G	G	F	G
R	Red	R	RH	4	M	R5
W	White	W	W	W	H	4
Y	Yellow	Y	Y	Y	C	Y6

Controllers

Analog controllers are used in many plants. They are hardwired with continuous outputs. Microprocessor-based controllers generally use a 0.05 to 0.5 second time cycle. Analog controllers use a standard signal level of 4 to 20 mA DC. Table 5-3 summarizes the types of controllers available and their basic characteristics. Table 5-4 lists a number of manufacturers that produce controllers.

Table 5-3 Controllers

Type Characteristics

A1	HVAC quality, 1/4, 1/8 DIN, RTD, TC, mV, mA DC service. Analog or microprocessor-based.
A2	Process control quality, electronic analog controller, 4 to 20 mA DC service.
B	Digital, microprocessor-based, configurable controller

Input/Output

A	1 to 5, 4 to 20 (standard), 10 to 50 mA DC, 1 to 5 V DC, 0 to 10 V DC, +/-2 mA DC, +/-10 V DC
B	Binary, binary-coded decimal, ASCII data per RS232, RS422, RS485, and IEEE488, TC and resistance temperature detector (RTD) inputs, analog I/O

Repeatability

A	0.25 to 0.5%, A2 has better repeatability
B	0.1 to 0.5%

Accuracy

A	1%, A2 better than A1
B	0.1% on input, TC and RTD 1°F or 0.5°C, setpoint 2%, outputs to valves 0.5%.

Control Modes

A	Manual on-off, proportional, integral, PL, PD, PID
B	Same as analog plus error squared, sample and hold, ratio, linearization, arithmetical and logical operations, counter-timer-selector or limiter functions, external feedback, feedforward compensation including lead/lag, and many other software options

Control Mode Adjustment Ranges (Standard)

A	Gain = 0.1 to 50, proportional band (PB) = 2% to 1000%, integral = 0.04 to 100 repeats/minute, derivative = 0.01 to 20-minutes
B	Gain = 0 to 128, PB = 0.8% to unlimited, integral = 0 to 293 repeats/minute, derivative = 0 to 895-minutes

Displays

A	Analog indications of process variable, setpoint, output, deviation, and balance.
B	Hybrid display of above data in digital and bar graph form plus a variety of status displays including lights and letter code abbreviations.

Table 5-4 Controller Manufacturers

ABB Kent-Taylor Inc. (A2,B)	Love Controls Corp. (A1)
Action Instruments (A1)	Moore Products Co. (A2,B)
Adatek Inc. (B)	Ohkiura Electric Co., Ltd. (A1)
Advanced Process Controls Inc. (B)	Omega Engineering (A1)
Air Monitor Corp. (A1)	Powell Process Systems (B)
Analogic Corp. (B)	Powers Process Controls (A1)
Athena Controls Inc. (A1)	Process Automation Co. (B)
Bailey Controls Co. (A2.b)	Robertshaw Controls Co. (A1)
Barber-Colman Co. (A1)	Rosemount Inc. (A2,B)
Beckman Industrial (A1)	Schlumberger Industries Inc. (A2)
Bristol Babcock Inc. (B)	Siemens Energy & Automation (B)
Burling Instruments Inc. (A1)	Smar International Corp. (B)
Chromlox Instruments & Controls (A1,B)	Syscon International, RKC Inst. Div. (A1)
Devar Inc. (A2)	Thermo Electric Co., Inc. (A1)
Dowty Custom Electronics Inc. (A1)	Tokhelm Corp. (B)
Drexelbrook Engineering (A1)	Toshiba International (A2,B)
Enercorp Instruments Ltd. (A1)	Triplett Corp. (A1)
Fischer & Porter Co. (A2,B)	United Electric Controls Co. (A1)
Fisher Controls International Inc. (B)	VTI Inc. (A1)
Foxboro Co. (A2,B)	Watlow Electric (A1)
Hartmann & Braun (B)	West Instruments (A1)
Honeywell Inc. (A1,A2,B)	Westinghouse Electric Corp. (A2,B)
Johnson Yokogawa Corp. (B)	Wilkerson Instrument Co. (A1)
Jumo Process Control Inc. (A1)	Yellow Springs Instrument Co. (A1)
Leeds & Northrup, a Unit of General Signal (A2)	Yokogawa Corp. (A2)
LFE Instruments (A1,B)	

The advantages of digital controllers include their accuracy and stability which are almost unlimited. Error sources such as hysteresis, nonlinearity, or thermal drift do not exist in a digital system. Digital systems are capable of sophisticated calculations and algorithms without recalibration and they can be easily and quickly reconfigured without requiring any change in hardware. They can provide more information and transmit data to other digital devices faster and more accurately. Digital controllers do not suffer from drift.

Any device that serves to maintain a variable value at a setpoint can be called a controller. The controller looks at a signal that represents the actual value of the variable, compares this signal to the setpoint, and acts on the process to minimize any difference between these two signals.

A simple control loop (Figure 5-3) contains a sensor, transmitter, control element and controller. The sensor measures the actual value of the variable. The transmitter amplifies this sense signal and transforms it into a form suitable for sending to the controller. The controller has two inputs: the measured signal and a setpoint signal. The setpoint may also be internally generated. The controller subtracts the two input signals producing a deviation or error signal. The controller may also reshape the deviation signal into an output according to the control requirements. The controller output then typically sets the operation of the final element in a direction to decrease the error.

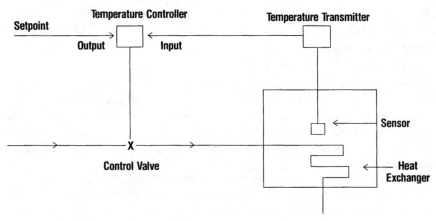

Figure 5-3 Temperature Control Loop

Controllers may be implemented in many different ways. The controller may work on pneumatic, fluidic, electric, magnetic, mechanical, or electronic principles, or on combinations of these.

Analog Controllers

An electronic controller has six basic parts or sections: input, control, output, display, switching, and power supply (Figure 5-4). The input section may generate the setpoint signal, condition the variable signal, and compare the two signals to produce a deviation signal. The control section generally uses a chopper-stabilized AC amplifier for the deviation signal. Other circuitry may amplify, integrate, and differentiate the deviation signal to produce an output with the necessary proportional gain, integral (reset), or derivative (rate) control actions. In some controllers the derivative circuit acts on the input signal rather than the deviation signal.

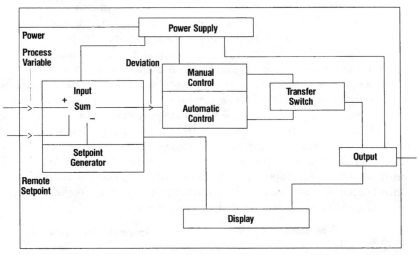

Figure 5-4 Basic Analog Controller Components

The controller generally has two modes: automatic and manual. A small display on the front panel provides information on the setpoint, variable value, deviation, and controller output. The front panel also allows switching between manual and automatic, and adjusting the value of the setpoint input and the manual output. Another indication, balance, allows the operator to equalize the manual and automatic outputs before transferring from the automatic to the manual mode.

The power supply transforms the incoming AC line voltage to the proper DC levels to operate the controller functions. Some controller power supplies also serve to power the transmitter in the same loop. In other cases the power supply may be external to the controller and used for energizing other controllers and instruments. Other features of the controller include alarms, feedforward inputs and output limits.

Alarms and Output Limiting

Common features involve alarm modules and output limits in some form. Electronic alarms can be setup on either the measurement, the deviation, or both. They can be actuated on the high or low side of the alarm point. There are alarm contacts for external alarms such as lights, bells, horns, or buzzers.

Most electronic controllers limit the high side of the output current range to the maximum standard value. A zener diode clamps the current at this value.

Some units place a fixed limit on the low end of the range. Others offer adjustable output limits on both ends.

Reset Windup

Limits can prevent reset (integral) windup which occurs if the deviation persists longer than it takes for the integral mode to drive the control amplifier to saturation. This can occur when there is a large disturbance or setpoint change. When the deviation changes sign (variable value crosses setpoint), the integral action reverses direction. So the variable is likely to overshoot the set-point by a large margin.

If the limit acts in the feedback section of the control amplifier's integral circuit, the controller output will go in the opposite direction as the input signal crosses the setpoint. This is called anti-reset windup.

Another controller feature is feedforward input. In these controllers a feedforward input signal is added to the controller output signal. Feedforward control (Figure 5-5) relieves the controller of compensating for a disturbance by normal feedback control. There are gain and bias modules for conditioning the feedforward signal before it sums with the controller output signal. A direct and reverse switch is used to match the feedforward signal to the controller action.

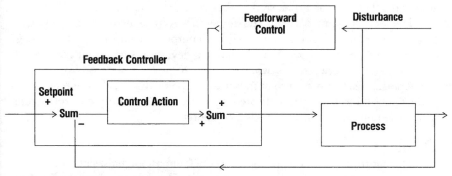

Figure 5-5 Feedforward Control

Some units have a communication jack that allows signal wires between the controller and transmitter to serve as an audio communication link. An AC carrier system is used that does not disrupt the DC process variable signal. This aids calibration and troubleshooting of equipment.

Trend recording can be done with jacks or pins for patching a recorder to the variable. Output tracking is used to lock the controller output to a remote signal. This is useful when an electronic controller backs up a computer control installation.

Input Signals

Input and output signals for electronic controllers are somewhat standardized, with three ranges:

- 1 to 5 mA DC,
- 4 to 20 mA DC and
- 10 to 50 mA DC.

The mid-range of 4 to 20 mA is the most common. The most common output is also 4 to 20 mA DC. Using a DC current amplifier for transmitting signals has one primary advantage over a voltage signal. Small resistances that develop in the line from worn switches and loose terminal connections do not change the signal value.

When the input signal reaches the controller, it is shunted across a resistor to produce a voltage that matches the controller's amplifier. The signals live zero helps with troubleshooting since it differentiates between the real signal zero and a shorted or grounded conductor.

Other signals are used such as 1 to 5 and 0 to 10 V DC, centered-zero signals and +/-10 V DC and +/-mA DC. Direct thermocouple (TC), resistance temperature detector (RTD) and mV signals are also available.

Setpoint Input

The controller's setpoint input can be internally (local) or externally (remote) generated. The two signals can be nulled or the unused signal can automatically track the other, so that the two signals are always equal.

The local setpoint is usually the output of a voltage divider that is connected to the setpoint adjustment. The divider's output is compared to the voltage drop produced by the input signal, yielding a deviation signal that corresponds to the difference between the two.

Another feature is a servo or motorized setpoint that tracks the remote setpoint value. Motorized setpoints are often used in supervisory control systems where the remote setpoint source is a computer.

Displays

The controller's front panel indicates the four basic signals: setpoint, process variable, error deviation, and controller output. The deviation meter sometimes doubles as a balance meter for equalizing the manual and automatic signals or local and remote setpoint signals when the operator wants to switch from automatic to manual output. There is also a meter to show the controller output. Controllers that have no panel displays are called blind controllers.

Most controllers merge the setpoint, deviation, and process variable indications into one display. This display consists of a deviation meter movement combined with a long steel tape or drum scale behind the meter's pointer. Only a portion of this scale shows. The tape or drum moves so that the scale value corresponding to the setpoint rests at the center of the display marked by a hairline.

The deviation meter movement corresponds to the visible part of the scale. When 20% of the scale shows, the deviation movement is +/-10% of full scale. The setpoint is either read out directly on a meter or displayed with a calibrated dial and index arrangement. The controller's output meter shows either the manual or the automatic output signal, depending on the transfer switch position.

Control Modes

The controller can deal a wide range of dynamic control characteristics. Adjustable control modes give it flexibility. These modes act on the deviation signal, producing changes in controller output to compensate for disturbances that have affected the controlled variable. Electronic modules can provide the following control actions:

- Proportional gain,
- Proportional gain plus integral (reset),
- Proportional gain plus derivative (rate),
- Proportional gain plus integral plus derivative and
- Integral (proportional speed floating).

The control action that is most widespread is proportional gain plus integral.

The proportional mode varies the controller output in proportion to changes in the deviation. High gain means that the controller output varies greatly for small changes in the deviation. Extremely high gains approach on-off control.

In systems with relatively low gains, the importance of the integral mode increases, because at low proportional gains the controller will be satisfied even though the controlled variable and the setpoint are relatively far apart (offset). The integral mode will continue to drive the output until the deviation goes to zero.

The derivative mode is a refinement that produces an output component proportional to the rate of change of either the deviation or the input signal. Its effect is to anticipate changes in the variable under control. Figure 5-6 shows a simple circuit for implementing three-mode control.

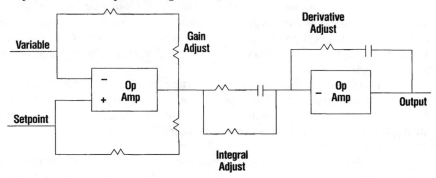

Figure 5-6 Three-mode Control

The advantage of having the derivative mode act on the input rather than on the deviation is that the controller does not respond directly to a setpoint change.

The adjustment of the control-mode lets the user fit or tune the controller to the system dynamics. The knobs are calibrated in terms of the degree of control-mode action. Typical adjustment values for proportional gain are 0.1 to 50, integral, 0.04 to 100 repeats per minute and derivative 0.01 to 20 minutes.

The proportional band (PB) is related inversely to proportional gain:

PB = $\dfrac{\textbf{100\%}}{\textbf{proportional gain}}$

One repeat per minute allows the integral mode to ramp the output up or down, repeating the proportional action that corresponds to a particular deviation every minute. Ramping continues until the deviation is zero. A derivative setting of one minute means that the controller output should satisfy a new set of operating conditions one minute sooner than it would without the derivative mode.

The direction on control action depends on the setpoint and process variable input connections. Direct control means that as the process variable increases, the controller output also increases. Reverse control means that the controller output responds in a direction opposite to that of the changing variable. A direct-reverse switch on the controller reverses the input connections and switches the controller action. If the derivative action is on the measurement rather than the deviation, this connection should be reversed also.

Some controllers allow changes in the control settings while the controller is operating in automatic without disturbing the output. Common load resistance values include 75, 500, 600, 800, 1500, and 3000 ohms.

Electronic Temperature Controllers

Besides the many general-purpose devices with standard inputs and outputs, analogous to the 3 to 15 PSIG (0.2 to 1.0 bar) standard signals in pneumatic control systems, a variety of other controllers are tailored to specific types of control. The most common types are electronic temperature controllers for driving electric heaters and motor speed controllers. These temperature controllers are designed to accept inputs from sensors such as thermocouples, resistance bulbs, or thermistors. Their outputs may drive relays, contactors or silicon-controlled rectifiers (SCRs).

Because the sensor connects directly to the controller, the input signal lines operate at low signal levels. The lines must be kept short to avoid interference from other sources. The controller's input circuit will compensate for variations in the thermocouples cold junction temperature. Resistance bulb and thermistor inputs generally feed into resistance bridge circuits.

The form of control ranges from simple two (on-off) or three-position (high-low-off) control to more sophisticated three-mode control. Some units offer three-mode control on one setpoint position and on-off control for another.

Proportioning is generally on a time basis called time-proportional control. The output is driven full on or off, but the control signal varies the percentage of on-to-off time in a duty cycle.

The average power level determines the load temperature. This type of control applies to both relay and SCR outputs. SCRs are capable of faster switching than relays, which allows proportioning times of a fraction of a second and smoother control.

Some controllers trigger the SCR to divide up each AC cycle while others proportion the number of AC cycles that are on and off in a certain period. SCRs can handle loads up to 300 kW using three line phases.

Position Controllers

One type of electronic controller with relay outputs is the position controller (Figure 5-7). This controller energizes one of two relays to operate an electric motor in either direction. The motor drives a final element to a desired position. A feedback signal from the final element tells the controller when the desired position has been reached. The controller acts as a master controller in a cascade system. The relays, motor, final element, and feedback signal make up the slave loop. The master controller generates the slave loop setpoint.

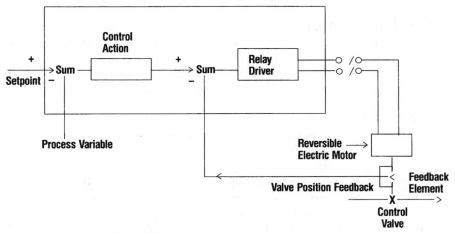

Figure 5-7 Valve Positioner Control

Nonlinear Controllers

This type of controller has an adjustable nonlinear relationship between measurement and output. The standard control modes of proportional, reset, and derivative are superimposed on the nonlinear function. For example, if the normal gain is set at 2 and the minimum gain within the dead zone is 0.2, then the effective gain inside that zone will be 0.04. This means that while the measurement is inside the dead band, a 25% change in measurement will result in only a 1% change in controller output due to the proportional mode response. The width of the dead band can be adjusted manually or automatically in direct or reverse proportion to an external signal.

A nonlinear controller can be used to filter out noise or pulsation while permitting effective control over major disturbances. In order to eliminate cycling during normal operation, the dead band should equal the band of noise or pulsation.

Balancing

In some controllers you can balance the two output signals before transferring from automatic to manual to avoid a sudden change in controller output that might disturb the process. This balanceless, bumpless transfer from manual to automatic output depends on a modified integral circuit to make the automatic signal track the manual output.

One method used for balancing switches the amplifier into a simple integrator configuration when the operator transfer from automatic to manual mode. A DC input signal is then sent to the integrator circuit input which causes the output to ramp up or down, depending on whether the input signal is positive or negative. The integral action eliminates any existing deviation between the setpoint and the variable values.

Servicing Features

Controller manufacturers use several approaches to aid troubleshooting. Manual and automatic sections are completely independent, plug-in modules. One can be removed without affecting the other. There may also be separate power supplies for testing. Some units have a plug-in unit for simulating a closed loop.

Optical Meter Relays

One way of implementing two- and three-position, high-gain control uses an optical meter relay. The input signal, such as a thermocouple, drives a galvanometer pointer to indicate temperature. Another pointer or index is manually set to the desired temperature on the same indicating scale. The relay or SCR is switched on when the two pointers are in the same position.

The setpoint arm has a small photocell and light source. When light strikes the photocell, the amplified current energizes the relay to supply power to a load. When the indicated temperature reaches the setpoint temperature, a vane on the indicating pointer breaks the light beam, and the relay is switched off. Some meter relays use two setpoint arms and relays for three-position control. The vane can be shaped to provide proportional control over a narrow temperature band.

Some meters use electronic amplifiers to boost the input signal. Other meter-relay packages use a resistance element connected as a voltage divider. It provides a voltage reference signal which is compared to the input.

Digital Controllers

A typical digital controller has a D/A and A/D converter, microprocessor, display and control sections and communications ports. (Figure 5-8) The microprocessor section contains the CPU, read-only memory (ROM) for program storage and the operating or random-access memory (RAM). The A/D converters on the input must have high resolution and high noise tolerance. This is usually obtained with voltage-to-frequency converters with counters or dual-slope integrators.

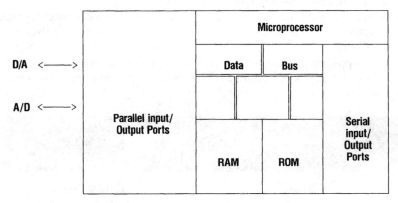

Figure 5-8 Digital Controller

Some controllers operate on pulse train inputs where the frequency indicates the value of the input. The D/A converters are usually implemented from resistor ladder networks.

The microprocessor calculates the setpoint, measurement, and output signals and operates the display. The cycle time varies from 0.05 to 0.5 seconds. Communications with other equipment are usually handled on an interrupt basis.

The program usually starts with reading the inputs, linearizing them, and checking them for alarms. This is followed by calculating the control algorithm. Then the displays are updated and output communications takes place.

Since the alarm points and tuning constants are stored in RAM memory, provisions are required to retain these values when the power supply fails. This is usually accomplished by the use of rechargeable batteries.

Communication formats include RS232, RS422, RS485 and IEEE488. The coding of the data also varies with manufacturers and includes ASCII, binary and binary coded decimal. Communication formats are often serial to minimize wiring requirements. They may be unidirectional, single wire, or bidirectional, two conductors. Data rates vary from kilobauds to several megabauds.

Software

Control functions and features include: lead-lag, feedforward, dead-time compensation, program pattern control, sample and hold, inverse derivative, gap control, auto select or limit functions, error squared, PID, and multiple inputs/outputs for cascade, ratio, or batch control. Many units can handle most of the different types of input signals. This includes low-level RTD and TC-type measurements with table look-up. Signals can be linearized, pressure-or temperature-compensated, converted to other engineering units totalized, time delayed, or manipulated arithmetically. Logic interlocks associated with the loop can also be implemented since many single-loop digital controller designs have both continuous and sequential control features. These include all the basic logic functions and are additional to the alarm features.

Programming of the controller is often done by sequentially entering the answers to a number of questions. You identify the engineering units of the controlled variable and indicate the reading to be considered as 0% reading. Next, the full span, upper range limit = 100%, would be identified, and then, the gain or proportional band, would be entered. The control algorithm and related features are also specified.

Table 5-5 Programming Parameters/Auto-Manual PID Controllers

Parameter	Typical Value
Engineering Units 0% Value	-99999 to 99999
Engineering Units 100% Value	-99999 to 99999
Gain	0 to 128
Reset	0 to 293 repeats/minute
Rate	0 to 895-minutes
Process Variable Filter Time Constant	0 to 1.8-minutes
Reverse Control Action	Yes or No
Increase Closes	Yes or No
Alarm Trip Points	-16 to 136% of span
Alarm Dead Band	0 to 136% of span
Setpoint High/Low Limits	-14 to 114% of span
Valve Output High/Low Limits	-14 to 114% of span
Anti-Reset Windup High/Low Limits	-14 to 114% of span
Feedforward Gain	0 to 1
Feedforward Reverse Action	Yes or No
Feedforward Filter Time Constant	0 to 112-minutes
Track Filter Time Constant	0 to 112-minutes
Restart Mode	1 to 5
Restart Valve Output	-14 to 114% of span
Restart Setpoint	-14 to 114% of span

In most cases, you can vary the basic algorithm; velocity, positional, sample-and-hold, error squared, noninteracting, external-reset, derivative and/or proportional acting on measurement and nonlinear. You can also provide linearization functions such as TC, RTD and counting and logic and safety interlock functions.

Some digital controllers also offer multivariable control or self-tuning/self-optimizing. Digital controllers provide much more flexibility for the particular application than analog units.

Single-loop controllers can be integrated into large systems using data bus communications. The data highway can also be used to provide a communication link to a plant-wide optimizing computer which can reset individual loops as required. These remote signals can modify setpoints, initiate auto/manual transferred, switch setpoints between local and remote sources and change controller actions. Some advanced units also provide self-calibrating, self-diagnostics, and self-tuning functions. Digital displays using light emitting diodes (LEDs), or liquid crystal bar graphs are used to convey the value of the variable.

The programming of some digital controllers can be performed using a keyboard which is an integral part of the controller. Others require separate plug-in programmers or are programmed through the keyboard of the central console.

Temperature Regulators

Temperature regulators are also known as temperature control valves (TCV). These units do not really offer the same control as a regular temperature control loop. Some TVCs can indicate the current measurement, but this is a special feature. They are also limited to proportional control and incapable of maintaining the setpoint without offset.

These are self-contained mechanical devices that provide proportional control. They are not capable of maintaining a setpoint because in order to change the output value, first the measured temperature has to change. All they can do is follow an operating line which relates the control element (load) to the controlled temperature (measurement).

The difference between the temperature setting and the actual process temperature is called the offset of the regulator. Although the resulting temperature control is not very accurate, these regulators are used in some HVAC applications.

The temperature is detected by a measuring element which senses the thermal expansion of the filling material. A number of substances are used in regulators. The regulator consists of the following parts:

- Detecting element (bulb),
- Measuring element (thermal actuator),
- Reference input (adjustment) and
- Final control element (valve).

Since they require no external power, regulators are self-actuated. They do take thermal energy from the controlled medium to obtain the required forces.

Regulator Design

Temperature regulators can be direct-actuated or pilot-actuated. In the direct-actuated type, the power unit is generally a bellows or diaphragm that is directly connected to the valve plug. The bellows or diagram develops the force and travel needed to fully open and close the valve.

In the pilot-actuated type, a thermal actuator moves a pilot valve. The pilot controls the pressure to a piston or diaphragm, which develops the thrust to position the valve. The pilot can be internal or external.

Direct-actuated regulators are generally simpler and lower in cost. They tend to be more proportional in their action with better stability. The pilot-actuated regulators have smaller bulbs with faster response. They have a narrower proportional band and they can handle higher pressures. Pilot-actuated regulators can be used for multiple functions such as temperature plus pressure.

Temperature regulators can also be self-contained or remote sensing. This depends on the location and structure of the measuring element (thermal actuator).

Self-contained regulators have the thermal actuator inside the valve body. The actuator serves as the primary detecting element so the self-contained units can only sense the temperature of the fluid flowing through the valve. This acts as both the controlling agent and the controlled medium since the device regulates the temperature of the fluid by regulating the fluids flow. Self-contained regulators typically use liquid expansion or fusion type thermal elements. The fusion type has a special wax filled with copper to enhance its thermal properties.

A common application of the internal sensor type of regulator is to keep the steam line full of steam while draining off only the condensate. An internal sensor type of temperature control valve (TCV) can be set at 200°F (93°C). It will open when it detects a lower temperature, releasing condensate, and will close when the temperature rises above that setting, preventing the release of steam.

In remote-sensing regulators, the bulb is separate from the power element of the thermal actuator. This style is able to sense and regulate the temperature of a fluid distinct from that of the fluid flowing through the valve. The self-contained style is simpler and lower in cost.

Thermal Actuators

The filled system thermal actuators used in temperature regulators are distinguished by the type of filling used. They all develop power and movement proportional to the measured temperature and are proportional controllers. Gas-filled and bimetal thermal elements are not normally used.

Vapor-filled systems are partially filled with a volatile liquid. The liquids used are chemically stable at temperatures well above the range used. The vapor pressure increases with rising temperature and acts on the bellows to move the valve plug.

Typical ranges of adjustment from −25 to 480°F (−31.7 to 248.9°C) are available with typical spans of 40 to 60°F (22 to 33°C). Special designs allow longer spans of up to 100°F (55.6°C) and shorter ranges of 25 to 30°F (13.9 to 16.7°C).

Refrigerant Control

One application is the control of heat exchangers which are cooled directly by refrigerant. The refrigerant control valve is operated by a diaphragm actuator which compares the pressure in the temperature bulb and the refrigerant vapor outlet pressure. The valve is balanced when bulb pressure equals the forces produced by the outlet pressure and the valve spring.

When the cooling load exceeds the refrigerant supply, the refrigerant vapor outlet temperature rises, causing an increase in the vapor pressure inside the bulb. This opens the valve and more refrigerant is evaporated. More cooling is done by the exchanger and the vapors become cooler so the balance is established.

If the cooling load drops, the refrigerant flow becomes excessive for the load and the temperature at the bulb drops. This closes the valve until the balance is established. The pressure drop variation across the evaporator tends to help in this operation.

Chapter 6
Wireless Control

Wireless control may be desirable for some types of HVAC systems. Remote control signals are usually transmitted over connecting wires and cables between the controller and the controlled device. In most HVAC applications, this will be the best choice, if only for economic reasons. In other applications a connecting cable may be inconvenient. Other alternatives exist and will be explored in this chapter. The most popular means of wireless control include:

- Light-beam control,
- Tone control,
- Carrier-current control and
- Radio control.

Each of these has its own set of advantages and disadvantages. There are also several possible variations in each of these categories.

Light-beam Control

In HVAC control you generally want to start out with electrical energy at the controller and end up with electrical energy at the controlled device. This energy can be converted to another form for transmission and then converted back to electrical energy.

Light is one form of energy that is relatively easy to convert back and forth from electrical energy. A light source such as an LED (light emitting diode) will emit light proportional to the electrical energy applied to it.

A photosensitive device is one that responds to light. Photosensitive electrical components are sometimes called photocells, but this usage can be confusing. A photocell actually refers to a photovoltaic cell or solar cell.

Photovoltaic Cells

Photovoltaic cells can be a simple PN junction, usually made of silicon. They are similar in construction to standard diodes, but the package allows the diode to be exposed to light. Silicon is a photosensitive material in a photovoltaic cell, the diode area is spread out into a relatively large, thin plate for the largest possible exposure to the light source.

The photovoltaic cell acts like a battery or dc voltage source while the light source supplies it with energy. Thus, the photovoltaic cell functions as a light-powered battery.

When the silicon surface is shielded from light, no current flows through the cell. When it is exposed to a bright light, a small voltage is generated within the cell because of the photoelectric effect. If an illuminated photovoltaic cell is hooked to a load, current flows through the circuit.

The amount of current that flows depends on the amount of light reaching the photosensitive surface of the cell. The brighter the light, the higher the current available from the cell.

The photovoltaic device's output voltage is relatively independent of the light level. The voltage produced by most commercially available photovoltaic cells is about 0.5 V.

This is the diode junction voltage. The 0.5 V output of a single cell is too low for most applications, so a number of photovoltaic cells are usually added together in series for more voltage and in parallel for more current.

These combinations of series and parallel connected photovoltaic cells provide a simple form of solar power and are used extensively in space applications. Solar batteries also provide power in remote areas for weather monitoring stations and other low power applications. The more cells there are in a solar battery, the larger the total surface area must be and the harder it is to arrange the cells so they will all be lighted evenly.

In a control system a photovoltaic cell can be used to trigger a sensitive relay. The controlled circuit will require a separate power supply since the photocell only opens and closes the relay contacts.

Since the output of a photovoltaic cell is relatively small, it can only be used to drive a low current relay or electronic switch. However, a light-duty relay or electronic switch can be used to control a larger, heavy-duty relay. A larger relay can also be driven by amplifying the output of a photovoltaic cell with a transistor amplifier circuit. These methods require an extra voltage source in addition to the photovoltaic cell and the controlled circuit's power supply. Since a photovoltaic cell's voltage output is relatively constant, this type of device cannot be used to monitor varying light levels.

Photoresistors

Another light-sensitive device is the photoresistor, or light-dependent resistor (LDR). The photoresistor changes its resistance value depending on the level of illumination on its surface. Photoresistors are junctionless devices like ordinary resistors and they have no fixed polarity. Photoresistors generate no voltage. They are usually made of cadmium sulfide or cadmium selenide.

These devices have a broad resistance range on the order of 10,000:1. The maximum resistance is about 1 meg-ohm (1,000,000 ohm) and is usually achieved when the cell is completely dark. As the light level increases, the resistance decreases.

Photoresistors are useful for HVAC control applications. They can be used to replace a variable resistor in a control circuit. Photoresistors can be used in many of the same applications as photovoltaic cells. They have the advantage of being sensitive to different light levels and can be used to trigger a control relay.

Other Photosensitive Devices

The other light-sensitive devices that are available include semiconductor devices, such as light-activated SCRs (LASCRs), photodiodes, and phototransistors. Phototransistors are useful for many applications because they can be used as amplifiers that are controlled by the light intensity.

An optoisolator is a device that isolates two interconnected circuits so that their only connection is optical (light). Most optoisolators consist of an LED and a photoresistor, photodiode, or phototransistor encapsulated in a single enclosed package.

The LED is wired into the controlling circuit and the phototransistor (or other photosensitive device) is wired into the circuit to be controlled. This provides a means of control with no electrical connection between the two circuits.

Light-controlled Relays

A light-controlled relay circuit can use a photoresistor and SCR which fires to drive a latching relay on. (Figure 6-1A) This could be used for a light-actuated fan control. The relay is activated when light shines on the photoresistor. The resistance drops as the cell's illumination increases. This allows the voltage across a neon lamp to rise. The lamp fires, triggering the SCR. The relays coil is connected across the SCR's output.

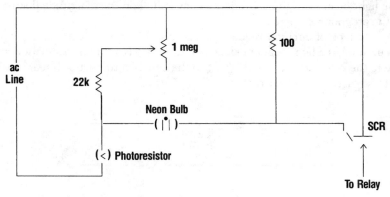

Figure 6-1a Light Controlled Circuits

A latching relay allows the light to be removed from the photoresistor without releasing the relays switching contacts. The controlled device can then be turned off by shining the light on the photoresistor a second time.

Another type of light-controlled relay circuit uses a sensitive operational amplifier (op amp) as a comparator. (Figure 6-1B) The op amp drives a transistor which in turn operates a relay to control a fan or other device. A photoresistor is used to trigger the op amp.

The op amp is set up as a comparator. The voltage dropped across the photoresistor is compared to the voltage taken off a potentiometer. When the comparator switches, it triggers the relay. The transistor is used to amplify the comparator's output signal. A low-power relay is not needed.

Besides the control circuits that activated by the presence of light, there are control circuits that can be used to detect the absence of light.

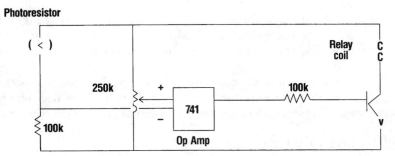

Figure 6-1b Light Controlled Circuits

Light-beam Transmitter

Some wireless applications involve modulating the light beam. Modulation is the process of imposing a programmed signal onto a carrier (the light beam). The light beam will dim and brighten depending on the amplitude fluctuations of the programmed signal.

The type of circuitry required to achieve this is not complex. One type of circuit used as a light-beam modulator/transmitter is a simple summation circuit where the programmed signal is sent to the base of a transistor. (Figure 6-2) An LED is placed in the emitter.

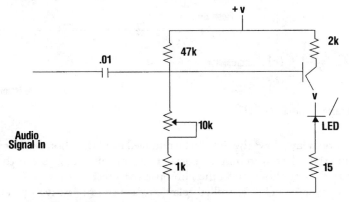

Figure 6-2 Light Beam Modulator/Transmitter

With no input signal, the LED is biased to glow at half brightness. When a changing signal is applied to the input, the LED's brightness will vary above and below this midpoint. A series resistor limits the current flow through the LED and an isolation capacitor prevents dc levels from affecting the input.

Light-beam Receivers

A light beam receiver/demodulator circuit is intended to pick up the signals sent out by the transmitter. The light from the LED in the transmitter reaches the solar cell of the receiver and is then amplified by a two stage RC-coupled, transistor amplifier.

The range of these circuits may be limited, but they are adequate for many remote control applications. The range will be better in a dim rather than a bright environment. External light sources can interfere with the light from the LED. An infrared emitter and sensor will work better, especially if ambient light levels are high. Another option is a fiberoptic connection, which increases the range and allows the signal to travel around corners.

Infrared Transmitters

A visible light-beam transmitter/receiver system can have some limitations. One problem is interference from external light sources. Also, in some applications, a visible beam of light might be undesirable.

A solution is to use an infrared light beam, which is invisible to the human eye. An infrared transmitter will use infrared LEDs. Multiple infrared LEDs in series will increase the range of the system. (Figure 6-3)

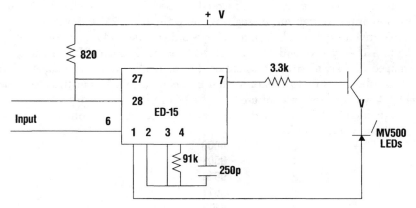

Figure 6-3 Infrared Transmitter

Infrared Receivers

An infrared receiver will use an array of infrared sensors. The system range can be as high as 30-feet. This is adequate for many remote control applications. These systems can only function along the line of sight. An infrared receiver circuit may use an op amp to amplify the sensor input. (Figure 6-4)

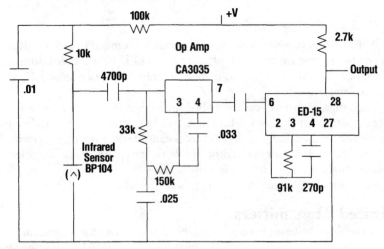

Figure 6-4 Infrared Receiver

Multifunction Infrared Transmitters

All of the light-actuated controls discussed so far can control only a single function. This is fine for turning a single device on and off.

In most cases a digital circuit would be used for encoded signals. A multifunction infrared transmitter such as those used for TV remote controls or remote inventory keypads requires a keyboard encoder (74922 or equivalent).

In a 16-key keyboard encoder, 16 NO (normally open) SPST switches are arranged in a x/y matrix. These are used to manually enter the data. A 16-key keypad allows up to 15 functions that can be controlled from these switches. The 0 key is ignored and not used.

An encoder IC converts the appropriate switch number into a four-bit digital value as shown:

0	0000
1	0001
2	0010
3	0011
4	0100
5	0101
6	0110
7	0111
8	1000
9	1001
A (10)	1010
B (11)	1011
C (12)	1100
D (13)	1101
E (14)	1110
F (15)	1111

This binary value is fed to the data inputs of a counter IC. When the borrow output of this chip goes high, it activates a multivibrator circuit that is connected to the count down input of another counter IC. This counter counts down from the loaded data value to zero.

On each counting pulse the output transistor is turned on, causing the infrared LEDs to flash. Multiple LEDs are used to increase the output level of the infrared signal for a maximum transmission range of about 15-feet.

The LEDs flash the number of times equal to the value of the closed switch. When the counter reaches zero, it cuts off the timer and waits for the next key to be depressed.

Multifunction Infrared Receivers

An infrared receiver can use an op amp such as the 748 along with a 74193 synchronous up/down counter and a phototransistor to detect the infrared pulses. These pulses are boosted by the high-gain amplifier and shaped into square waves. The squared pulses are sent to the counter IC. The counter counts the pulses and feeds the value in binary form through four digital outputs.

Chopped Light Control

In wireless control the activating light does not have to be in the form of a continuous beam. This is used in many applications, but it can cause problems in some applications. In some control circuits it is desirable to use chopped light. The chopped signal is a stream of regular pulses. The chopped light signal is not a continuous beam, but it is modulated (switched on and off) at a regular rate. In most applications, the chop frequency, which is the rate at which the light is switched on and off, will be in the mid-audio range, typically between about 400 Hz and 1 kHz (1000 Hz).

In a control circuit for a relay operated by a chopped light source the relay is activated by a chopped light beam. (Figure 6-5) The light beam is sensed by a photovoltaic cell which is capacitively coupled to an op amp IC. The output of the op amp is then capacitively coupled to an output relay.

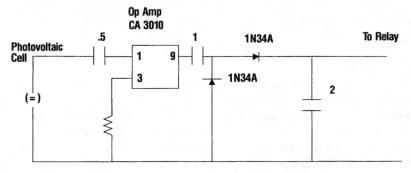

Figure 6-5 Chopped Light Controller

If the controlling light beams modulation frequency (chop rate) is too low, there can be a problem with relay chatter. The relays contacts can erratically open and close multiple times instead of neatly making or breaking contact. This usually means that the capacitor across the relay coil is too low or leaky.

Light-controlled Capacitance

Many electrical parameters can be controlled by a variable resistance. In these cases, a photoresistor is used in the control. In some applications a variable capacitance instead of a variable resistance may be used.

A varactor diode will often be used to control a capacitance via light intensity. The light-controlled capacitance circuit will use a photovoltaic cell as the light sensor. The varactor diode capacitance will vary in response to the light intensity shining on this sensor.

A varactor diode is a specialized two-lead semiconductor device that is designed to provide an electrically variable capacitance. All semiconductor diodes exhibit some amount of capacitance when they are reverse biased. In most diode applications this capacitance is usually undesirable since it limits the operating or switching frequency of the diode.

The varactor diode uses this capacitance as a controllable characteristic. The voltage applied to the reverse-biased varactor diode determines this capacitance.

Varactor diodes are used for frequency modulation (FM), frequency multipliers and tuners for radio frequency (RF) amplifiers. Since the varactor diode acts as a variable capacitance, it is also called a varicap.

A high value isolating resistor is usually needed in varactor diode applications. Many times several resistors in the meg-ohm range are used. A single higher-valued resistor can exhibit stability problems.

A capacitance is used to block any dc voltage from the photocell from reaching the circuit's output terminals. This capacitor also protects the varactor by making it impossible for an external load circuit to place a damaging short circuit across the varactor diode. This capacitor's value is much higher than the maximum varactor capacitance, so it is a short circuit at most operating frequencies. This type of light-controlled capacitance circuit can be used to control an oscillator.

Fiberoptics

Light-based remote control systems tend to have relatively limited transmission ranges. They are also limited to line-of-sight systems. If any object gets between the light transmitter and its receiver, the signal will be blocked.

One solution is to use a fiberoptic cable. In a fiberoptic system, special light conducting cables are used between the transmitter to the receiver. Unlike ordinary wires, the fiberoptic cables carry light pulses.

The fiberoptic cable is made up of multiple hair-thin glass or plastic fibers. These cables are flexible and can be bent to conduct the light signal to a desired destination. The light at the output is subject to attenuation affects depending on the length of the cable.

However, if you use a fiberoptic cable, you lose the wireless advantage. But, fiberoptic cables offer several advantages over ordinary electrical wiring. Because no electrical signal is carried through the cable, it is safer. There is no shock or fire hazard because of worn insulation or short circuits.

Optical fibers are commonly used in communications because of their long distance high bandwidth capabilities. While these characteristics are valuable in most applications, the overwhelming reason for using fiberoptics in industrial communications is their immunity to electromagnetic interference (EMI).

Many plants use large electric motors to operate conveyors and pumps. These large electrical loads create electromagnetic waves and ground surges. Plants that are composed of several structures usually have separate power distribution systems, resulting in differences in ground potential.

A fiberoptic cable of a given thickness can also carry more independent signals than a comparably sized electrical cable. Telephone lines are using more fiberoptics for these reasons.

Fiberoptic cables tend to be lighter and more flexible than comparable electrical cables. If a bend in an electrical cable is too sharp, an internal break in a wire can result. Fiberoptic cable is more expensive than ordinary electrical wire, so it is only used in applications in which its special advantages are important.

Fiberoptics Conductors

The center of the fiber cable is called the core. It conducts the light, similar to the conductor in a wire. The outer part of the fiber is called the cladding. It keeps the light in the core. Outside the cladding is a protective polymeric coating to give the fiber mechanical strength.

Both the cladding and the core are transparent. The cladding has a lower index of refraction than the core, so the light will be reflected by the cladding back into the core. Optical fibers are subject to sources of attenuation such as macrobends and dirt.

There are many standard sizes of optical fibers. The sizes are specified by the diameter of the core and the diameter of cladding in microns.

Fibers with smaller cores tend to have smaller attenuations and larger bandwidths, which are required for long-distance applications. Fiber with larger cores tend to have greater tolerance to dirt or bending, and faster connection installation. See Table 6-1.

Step index fibers have a constant index of refraction from the center of the core to the edge of the cladding and then a single step down to the index of refraction of the cladding. Graded index fibers have a smooth downward gradation of index of refraction from the center of the core to the edge of the cladding. Step index fibers collect twice as much light as graded index.

Table 6-1 Fiber Characteristics

Fiber size	9/125	62.5/125	220/230
Longest distance between repeaters	> 3 km	< 3 km	< 2 km
Greatest bandwidth at longest distance	> 1 GHz	200 MHz	< 10 MHz
Environmental tolerance	very clean	clean	dirty
Installation and maintenance level	expert	specialist	electrician

Fiberoptic Transmitters

A remote control transmitter for fiberoptic transmission is similar to those already discussed. The remote transmitter circuit output is designed with an optical port for transferring the LED light output to the fiberoptic cable. An op amp and transistor are used to control the LED.

The modulated signal in this type of circuit can be as high as several MHz (1,000,000 Hz). When no input signal is applied, the LED current will be about 50 mA (0.05 A). The current across the LED will vary from 0 to 100 mA (0.1 A) with the ac modulating signal.

Transmitters are similar in construction to detectors, but an emitter integrated circuit package replaces the detector IC. The light from a typical LED is imaged into a spot 100 to 250 microns in diameters.

In large-core fibers the core is almost as big as the spot of light at the window of the LED, so very little light spills over into the cladding and is lost.

Fiberoptic Receivers

A fiberoptic receiver will use an optical port to transfer the light to a photodiode. The sensitivity of the photodiode in these circuits is about 0.5 A/W. An op amp boosts this small signal to a usable level which is passed on to the output. This system could be used for on/off control, or a modulator and demodulator can be added for multichannel control. A typical op amp that is used is the LH032 op amp and a common photodiode is the HP5082-4220.

Receivers generally use integrated circuit packages containing both the light detector and amplifier circuit. Light comes into the detector and is amplified and sent through a lens for focusing onto the optical fiber. The detectors are usually semiconductor diodes, either PIN or avalanche design. Avalanche diodes are normally used with 9/125 fiber. PIN diodes are used with 62.5/125 and 200/230 fiber due to their lower cost, wider temperature range, and greater stability.

Light sources for optical fibers are almost exclusively semiconductor light emitting diodes and lasers. Lasers are commonly used with 9/125 fiber. Light emitting diodes (LEDs) are used with 62.5/125 and 200/230 fiber for their lower cost, wider temperature range and longer lifetimes.

Fiberoptic Connectors

Among the major connector designs, the SMA, ST, and SC types are commonly used in data communications. The biconic, FC, and D4 connectors were designed for long distance telephony. Table 6-2 lists the losses for the different types of connectors.

Table 6-2 Connector Losses

		SMA	ST	SC
Loss from 10 micron particle on ferrule side				
62.5/125 fiber	dB	0.35	0.70	1.00
Loss from 20 micron particle on fiber core				
62.5/125 fiber	dB	1.03	1.03	1.03

The typical method of mounting connectors on the optical cables uses adhesives and polishing. These techniques require specialized training and experience to provide reasonable installation times.

Each of these connector designs has advantages and disadvantages. The SMA connector was designed for dirty environments. For 200/230 fiber, the combination of large core and thin cladding allows the use of connectors that can be installed with simple tools. Their crimp and cleave operation requires no adhesive and no polishing. It take less than five minutes to install a connector, including the time to prepare the cable. All of the installation operations are hand-held and do not require work space to lay out equipment. The 200/230 fiber was designed for factory automation applications.

Because of its low connector cost, the most popular type of fiberoptic connector is the epoxy-polish (EP) type requiring two-part epoxy, heat curing and polishing. This type is best installed in a shop atmosphere by personnel who are skilled in producing many cable assemblies in a controlled atmosphere. In the field the EP type is the least desirable because of the setup and tear-down time required, working space requirements, 115 VAC power requirement, consumables and scrap costs. Similar to the EP type, the UV-cure uses a one-part epoxy and UV-curing lamp.

There is also a no-epoxy/crimp-type connector, but this no-epoxy-crimp connector must be polished just like the epoxy polish and UV cure types.

The most important operation for terminating any of these connectors is the polishing, which is done in two or more steps. Improper polishing leaves a poor finish such as scratches and pits which degrade the light signal. The polished fiber should also be inspected under a microscope. Sometimes very small scratches may be acceptable and testing with a power meter can determine acceptability. Pits in the core region, large scratches and fractures are unacceptable and if these imperfections cannot be eliminated, then a new connector must be installed.

A popular connector for field use is a no-epoxy/no-polish type made by Siecor known as the Universal CamLite (UniCam™). Although the cost of the connector is higher than the epoxy-polish type, significant installation time is saved. Since it is easy to install, it can provide a good connection each time, even by an inexperienced installer.

Fiberoptic Cable

The two basic types of fiberoptic cable constructions are the buffer and loose tube. Indoor and short-distance outdoor applications use a tight buffer construction. This type of construction supports the fiber and has greater crush and impact resistance for cable tray applications. It usually has strength members for each fiber and the connectors can be attached directly. Tight buffer construction cables are available for flame-retardant, aerial, and cable tray installations.

Long-distance outdoor applications normally use a loose tube construction. In the loose tube construction, the fiber is loosely laid inside a plastic tube and filled with a water-repellent gel. This construction insulates the fiber from stresses on the cable such as thermal expansion and contraction. It usually has less attenuation in outdoor, long-distance applications. Loose tube construction cables can be obtained for aerial, conduit, and buried installations. This construction lacks strength members on each fiber and requires accessories to attach connectors.

Fiberoptic Networks

Fiberoptic networks are becoming more common in factory automation and process control applications. These networks are a utility, similar to the electric power or telephone system. The network must support any connected device and support standard industrial communication protocols.

Fiberoptic Modems

A modem translates the communication signals from one type of device into the signals for another type of device. The most common form of modem converts the digital electrical RS-232 communication signals from a personal computer into the analog electrical signals required for the telephone network. These modems allow the use of the telephone network for long distance communications between computers.

In a fiberoptic modem, the operation is similar. In industrial communications, digital computers, programmable logic controllers (PLCs) and their remote I/O drops may need to communicate over fiberoptic cable. The fiberoptic modem converts digital electrical signals into digital optical signals to be transmitted by optical fibers.

Fiber modems may be used to extend the range or provide electrical immunity. For protection against transients, fiberoptics between locations provide total electrical isolation. This same capability can be used to protect against systems with poor grounding, eliminating ground-loop paths in the system.

Wide ranges of data rates, 1,200 to 115,200-baud asynchronous and from 1- to 100-Mbaud synchronous are available. The distance for communications over fiber can range from a few feet to several miles. The range can be extended further with repeaters. Ethernet can also operate on fiberoptic modems. Ethernet communication on fiber networks has been accomplished for years.

Fiberoptic modems can be configured to operate with redundant self-healing paths, which can automatically reroute the data flow if a break in the fiber path or an equipment failure occurs.

PLC Modems

A PLC modem converts each electrical pulse into an equivalent optical pulse. There is no sampling or multiplexing of the pulse, so there is no change in the pulse-to-pulse timing. The attached PLCs and remote I/O drops operate as if they were linked by ordinary cabling. The main difference is that they will experience no electrical noise from radio frequency interference, ground loops, lightning surges, and other sources.

PLC Optical Systems

Programmable logic controllers are a common type of digital control system in industrial applications. A typical application is one master controller with several remote input/output (slave) modules.

One configuration of the remote input/output modules is a linear string or daisy chain. When the daisy chain is electrical, each remote I/O module taps onto the copper cable. When it is a fiberoptic system, each remote I/O repeats the signal. A remote I/O string of 20 units would have 19 repeaters.

Pulse repetition is an important aspect of the digital signal transmission. Pulse width distortion limits the number of repeaters that can be used. A chain of repeaters can distort the pulse sufficiently to cause communication errors. Many PLCs can operate with 20% pulse width distortion. If each repeater adds 1% distortion, the maximum number of repeaters is 20.

Until recently PLC modems were designed as single-purpose devices. With the increased use of optical fiber networks and advances in fiberoptic technology, PLC modem requirements have changed. Modems support all the standard PLCs even though some of the signaling methods are proprietary.

Carrier-current Control Systems

It is sometimes convenient to use the ac electrical wiring that is already installed in most buildings. These ac-powered systems are not really wireless because the controller and controlled device are connected to the ac wiring. There is a physical connection between the two units.

Modulation of the 60-Hz ac power line signal can be accomplished with a control signal. This is called carrier-current control.

There are some limitations to this system and this is why it is not universally used. There must be isolation of any part of the circuitry the user might come in contact with from the ac line. Optoisolators are often utilized for this isolation.

Also, ac house wiring was never designed for communication of modulated signals. Noise and interference can be so severe that the carrier-current system becomes useless. These problems may or may not be temporary. It depends on the demands of the same ac power lines.

The convenience of carrier-current control makes it a viable alternative to consider in a HVAC control system. A typical carrier-current control system will have a control box or controller that plugs into the ac line. At another location the receiver/demodulator and load are plugged in. The ac line acts as the connection for signals between the remote devices.

In an actual ac power system there are usually several separate circuits. The service entrance is usually made up of at least three wires. This would be a single phase system. One of these is a common line and the voltage between it and either of the other wires is nominally 120-V ac. The actual voltage will depend on the load on the system which includes the equipment used by the power company. The voltage between the two outer lines will be about 240-V ac.

In a carrier-current remote control system the transmitter can be plugged into one line and the receiver can be plugged into the other. There may not always be a direct connection available. In larger buildings with several distribution boxes there may be several circuits between the transmitter and receiver. Another potential problem is that you cannot control what other loads are plugged into the circuit when the system is in use.

Carrier-current Signal Frequency

Many of these problems are handled by using the proper frequency for the carrier-current signal. It is also possible to condition the wiring so that it will carry the control signals without degradation.

A high-frequency (RF) transmission line should be terminated in its characteristic impedance, or else standing waves may be created along the length of the line. Standing waves mean that the voltage and current will vary along the length of the line. For the lower-frequency signals, this is not a problem and termination and the line length are not as important as with high-frequency signals.

The ac line acts like a random network of various resistances, capacitances, and inductances. There is no way to predict its exact electrical characteristics so it is better to use a low frequency signal.

The ac power lines in the United States already carry a very strong 60 Hz signal. This means you should not use a signal close to 60 Hz. It would just get lost in the power signal. Also, you need to avoid any harmonics of 60 Hz (120 Hz, 180 Hz, 240 Hz, 300 Hz, 360 Hz, etc.). Low frequencies in this range would have to be extremely strong to overcome the 60 Hz noise.

The carrier-current signal that is normally used is in the 60-to-180-kHz range. This is high enough to reduce the harmonics from the power line to an insignificant level, but low enough so that RF interference and transmission line problems should be small.

Tone Encoding

In a carrier-current system the control signals are tone encoded. These encoded tones are used to modulate the carrier signal. Both amplitude modulation (AM) or frequency modulation (FM) are used. FM systems are more complex and expensive, but tend to be less noisy. The tone encoding frequencies should also avoid 60 Hz and its harmonics, to limit interference from the power line. Many systems adapt the dual-tone frequencies used in Touch Tone telephones.

Filtering

To couple the tone signal into the ac line you must filter out the 60 Hz signal and possibly some of its lower harmonics. This is done with a narrow band-reject or notch filter. The simplest type is called a twin-T network or parallel T since it resembles two Ts. (Figure 6-6) One T is made of two similar capacitors in the upper end and a resistance in the lower part. The other T has two similar resistors on the top branches and a capacitor in the lower section. The Ts are connected in parallel and the values must be selected to center the rejection band around 60 Hz.

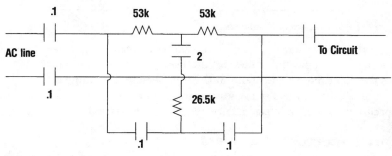

Figure 6-6 Twin-t coupling network

Carrier-current Transmitters

A carrier-current transmitter is usually made up of three primary stages:

- Carrier frequency generator,
- Transmitter and
- Coupling network.

The typical carrier frequency generator circuit is a (voltage-controlled oscillator) VCO. A 555 timer IC may be used as a VCO. The input signal modulates the output frequency using frequency modulation (FM).

The output frequencies will range from about 1 kHz to over 100 kHz. The input frequencies must be significantly lower than the carrier frequency to avoid aliasing problems.

The coupling network is a twin-T filter, as discussed earlier, with a few coupling capacitors. The carrier-current transmitter circuit can be a simple two transistor driver. (Figure 6-7) A tunable coil is used to provide the maximum drive for the transistor driver. The coil is similar to a TV width coil. This circuit presents a very low impedance to the ac power line and can be used with either AM or FM signals.

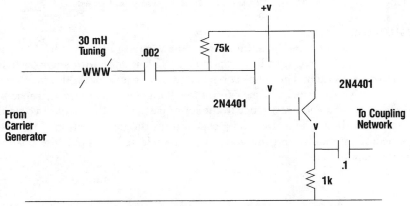

Figure 6-7 Carrier-Current transmitter

If the carrier-current signal is transmitted over a long distance in a large building, more powerful transistors are used. Another variation of this circuit uses the 555 timer to precondition the input before the tuning coil.

Other carrier-current transmitter circuits use a variable capacitor for tuning. The capacitor is shunted by a fixed inductor similar to a ferrite-rod AM radio antenna.

Carrier-current Receivers

The receiver circuit can be a simple R-C coupled amplifier which goes between the coupling network and the detector for the tone decoder. (Figure 6-8) Reception is improved with an impedance matching network at the input. Another type of carrier-current receiver circuit uses a single PNP transistor with a ferrite-rod shunted by a variable capacitor for tuning.

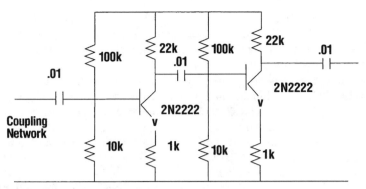

Figure 6-8 Carrier-current receiver

Tone Decoders

A typical tone decoder circuit will use the 567 tone decoder IC. This device is manufactured for this type of application. Some types of carrier-current receiver circuits use a built-in tone decoder. The 567 tone decoder IC is used with five capacitors and three resistors.

Tone Encoding Control

Using different frequencies for different functions is one way to combine multiple control signals on a single channel. Almost any transmission method can be used, including direct connection through wires, RF signals, optical signals, or ac carrier currents.

Another transmission method is sonic. The sounds can be sent with a speaker or sonic transducer and detected with a microphone. Audio control tones can be annoying or distracting to workers so ultrasonic signals above 20 kHz are used. These will function in much the same way, but they are not audible to humans. Ultrasonic transducers (speakers and microphones) are available and similar to high frequency audio transducers.

Simple on/off control can be accomplished with a burst of sound and a VOX (voice-operated switch) circuit. However, false triggering because of environmental noise can become a problem.

Better performance can be achieved by using a specific frequency for the control function. The controlled device will recognize only its assigned frequency. Sounds at other frequencies will be ignored.

Another frequency can be assigned to a second control function and yet another to a third function. This is tone encoding where the tones are generated by a tone encoder. In such a tone-operated system, the output of each tone decoder can be a dc control signal. This type of system allows multiple functions to be controlled from a single channel.

Tone encoding is usually done with one or more oscillator circuits. The output is gated on and off to activate the control function.

Tone decoding is more complex. The two basic approaches involve bandpass filters or phase-locked loops (PLLs).

Filter Decoding

In the filter method, a band-pass filter that passes only those frequencies which lie within a specific band or range is used. Signals with frequencies outside this band will not reach the output.

An ideal band-pass filter rejects all other frequency signals that are outside its pass band. Actual filter circuits cannot distinguish quite so well between frequencies and there is some leakage with a more gradual change between the pass band and the rejection bands.

The band-pass filter has a center frequency which is the midpoint of the passed band. The bandwidth (BW) is the width of the passed band. It indicates those frequencies that are passed.

If a band-pass filter passes only those frequencies between 2000 and 4000 Hz, the center frequency is 3000 Hz, and the bandwidth is 4000 − 2000 = 2000 Hz.

Frequencies close to the pass band of a practical filter will get through to the output, but they will be greatly attenuated. The more ideal the filter is, the better its selectivity will be. Highly selective filters use op amp circuits and are known as active filters.

In a multichannel filter decoding receiver, several separate band-pass filters are used, one for each encoded frequency that is used for a controlled function. The control signal triggers the appropriate device if and only if there is an output from the related filter. This only happens when the input frequency is within that filters pass band.

PLL Decoding

The phase-locked loop (PLL) is a type of closed-loop control system with feedback. There are three critical parts:

- Phase comparator,
- Low-pass filter/amplifier and
- VCO (voltage-controlled oscillator).

All three of these are generally on a single IC chip. A PLL such as the 567 has all these components and is designed for tone decoding applications.

A basic tone decoder built with the 567 uses four capacitors and a resistor. One resistor and capacitor set the free-running (center) frequency of the current-controlled oscillator. This will be the frequency detected by the tone detector.

PLL Operation

Two signals are fed into the phase comparator. One is the external input signal and the other is the output of the VCO, which is the feedback loop.

If these two signals are equal and 90 degrees out of phase, there will be no output from the phase comparator. The output of the PLL, which is the VCO's frequency, will not change.

When the two signals are not in lock-step, the phase comparator produces an output called the error signal. This signal is fed through the low-pass filter for smoothing and to prevent oscillations of the closed-loop system. The error signal is also amplified and fed to the control input of the VCO, changing its phase and possibly its frequency, until it is locked onto (and 90 degrees out of phase with) the external input signal.

Noise and Response Characteristics

Noise immunity for this type of tone decoder is achieved by adjusting the amount of time required for it to respond to a tone. A slower reaction time allows the circuit to ignore transients that might otherwise control it.

The response time is adjusted by changing the circuit's bandwidth. The wider the bandwidth, the quicker the response. A slower, more noise-immune response requires a narrow bandwidth.

The bandwidth is determined by the applied signal voltage, the center frequency, and a capacitor. The usable range of center frequencies for the 567 goes from 0.01 Hz to 500 kHz (500,000 Hz). This range can handle most control applications.

Touch-tone Encoding

When several control functions are needed in a tone encoding system, complex tones are used. Telephones use the Touch Tone system where each digit is represented by a combination of two tones. Both tones must be present for a valid control signal. This limits the chances of false triggering.

ICs for generating these signals can provide up to 12 control signals for a 3 X 4 column/row matrix keypad. When one of the key buttons is pressed, a low frequency and a high frequency are simultaneously generated.

One typical IC is the Motorola MC14410 tone encoder chip. The tones are derived from a 1-MHz crystal oscillator. This 1-MHz frequency is divided digitally to create each of the required frequencies. An additional column is supported by this device which allows a total of 16 control switches.

At the receiving end of the encoding circuit, a tone decoder for each of the component frequencies is provided. Gates are used to combine the tone decoder outputs.

Radio Control

Wireless control also includes radio control. While radio control might be desirable in some applications, it has several problems including interference and legal restrictions. Even with these problems, there are some applications where radio control may be the best or only practical control method. Radio frequency (RF) transmitters are under the jurisdiction of the Federal Communications Commission (FCC).

In most cases, radio transmitters must be licensed by the FCC to be legally used. The FCC also assigns specific functions to the various frequency bands in the radio spectrum. Three bands have been assigned to radio control devices:

- 27 MHz,
- 50 to 54 MHz and

- 72 to 76 MHz.

The highest (72 to 75 MHz) and the lowest (27 MHz) of these bands do not require an operator's license, but there are restrictions on their use. Each band is divided into several channels:

27 MHz band

26.995 MHz	27.045 MHz	27.095 MHz
27.195 MHz	27.225 MHz	27.145 MHz

72 to 76 MHz band

72.08 MHz	72.16 MHz
72.24 MHz	72.32 MHz
72.40 MHz	72.96 MHz
75.64 MHz	

The four frequencies in the first column of the 72 to 76 MHz band are restricted for use with model aircraft only.

An amateur radio technician class license (or better) is required for the 50 to 54 MHz band. The assigned channels in this band are

51.20 MHz	52.04 MHz	53.10 MHz
53.20 MHz	53.30 MHz	53.40 MHz
53.50 MHz		

Since many of the channel assignments in each band are closely spaced, highly selective receivers of the superheterodyne type must be used. Transmitters for devices such as garage door openers or radio-controlled model airplanes are certified by the FCC but it is illegal to make any modifications to the standard transmitters including a longer antenna. There are no legal restrictions for modifying a receiver.

Low-power AM Transmitters

Small AM transmitters are unlicensed for the AM broadcast band of 54 to 160 kHz. The transmitted signal is limited to a 50-foot range. An AM wireless microphone circuit only needs an op amp and two transistors. Instead of a microphone for voice input, control tones can be used. The signal may be picked up by a nearby AM radio that is tuned to the transmitter frequency. These signals can go to tone decoders and used as discussed earlier for carrier-current systems. One problem with operating in the low power AM mode is the considerable interference that can be encountered in this band which limits its usefulness for control applications.

Radio Modems

A radio modem provides a wireless Radio Area Network (RAN) interface and allows asynchronous RS-232C, RS-422, and RS-485 modules to communicate with PLCs and PCs. This allows user communication to inaccessible areas and eliminates in-plant wiring, leased line, and cellular communication costs.

Radio modems also eliminate shifting ground-plane and ground-loop problems and can be less expensive and more convenient to deploy than either fiber or wire-type modems. Rapidly advancing radio technology from military and space programs allows spread-spectrum radio modems for reliable data communications. Spread-spectrum modems are available with ranges from several feet to many miles and asynchronous data rates of up to 115,200 baud. Most spread-spectrum radio modems operate in the license-free ISM (industrial/scientific/medical) 902-to 928-MHz and 2,400 to 2,438-GHz frequency bands.

Frequency-hopping, spread-spectrum technology is particularly immune to interference. This technique packetizes the data stream and transmits the packets on a hundred or more sub-bands using a pseudo-random hopping pattern. Packets that might be interfered with are retransmitted at different frequencies using preset hopping algorithms. Frequency hopping allows the channels to operate in close proximity to each other allowing redundant channel operation.

Spread-spectrum radio modems can be operated as point-to-point, multipoint, or multidrop. Both polling and reporting by exception can be used. Repeaters can be used for range extension to overcome line-of-sight limitations and for redundancy.

Ethernet Radio Modems

Ethernet radio modems have emerged from spread-spectrum radio modems. Ethernet is a popular industrial communications medium because of its speed, ease of application and inherent data-multiplexing capabilities. They have characteristics similar to spread-spectrum radio modems. The 2.4-GHz band modems offer a throughput of 1-MBps and the 902-MHz band units offer 80-Kbps. The 902-MHz band has a greater range and is less susceptible to interference.

Ethernet spread-spectrum radio modems can be used for redundant operation. Both point-to-point and multipoint Ethernet network bridging can be done with dual radio modems on parallel paths.

The ABB Kent-Taylor ESTeem modem supports the Modcell MODBUS and ICN LINK communication protocols. This allows users to Read or Write memory between controllers. Once the modem is configured by MODBUS or ICN LINK protocol, the modem will read the destination address in the protocol and send the information to the modem and PLC with that address. The modem can route information through multiple modems to extend the range of the network transparent to the devices. Features include:

- Wireless networking communication protocols,
- Direct interface to controller via RS-232 and RS-422 interfaces,
- Point-to-point, point-to-multipoint, and polled with report by exception protocols,
- Functions as a master, remote, or repeater,
- Remote programmability,
- Field programmable VHF and UHF radio frequencies,
- Narrow band packet burst technology with CSMA-CD protocol,
- Radio self-test and packet monitor.

Wireless Networks

Arizona's Vail school district uses a wireless area network to connect its three schools and 15 buildings. BreezeCOM's BreezeNET radio transmitters and antennas are used. BreezeNET operates with a data rate of up to 3 megabits/second and is compatible with 802.3 Ethernet. It has a range of up to 8-miles outdoors, 600-feet indoors and roaming speeds to 60-mph.

Wireless LANs allow workstations to communicate and to access the network using radio propagation as the transmission medium. The wireless LAN can be connected to an existing wired LAN as an extension, or can form the basis of a new network. While adaptable to both indoor and outdoor environments, wireless LANs are more suited to indoor locations such as office buildings, manufacturing floors, hospitals and universities.

Cells

The basic building block of the wireless LAN is the cell. This is the area in which the wireless communication takes place. The coverage area of a cell depends upon the strength of the propagated radio signal and the type and construction of walls, partitions and other physical characteristics of the indoor environment. In general, a cell covers approximately a circular area. PC-based workstations, notebook and pen-based computers can move around freely in the cell. All radio communication in the cell is coordinated by a traffic management function. In the BreezeNet system, the radio traffic management function is performed by a unit called an access point. The access point connects the cells of the wireless LAN with one another and connected wireless LAN cells to a wired Ethernet LAN via a cable connection to an Ethernet LAN outlet.

A wireless LAN can be built up from standalone cells or cells connected to Ethernet. Overlapping cells can also be used including linked cells and multi-cells.

Standalone Cells

In a standalone cell system, the basic cell is made up of an access point and the associated wireless stations. The number of wireless stations per cell depends on the amount and type of data traffic. In a busy traffic environment a cell might contain 50 stations while in a slower environment 200 stations might be supported. The stations communicate with each other via the access point which manages the data traffic in the cell.

A standalone cell is a good method for setting up a small to medium-sized LAN between a number of workstations or workgroups. This type of cell requires no cabling.

Ethernet Cells

Cells can be connected to a wired Ethernet LAN via an access point and remote LANs via a wireless bridge. In a wired system the access point connects to the backbone of a wired Ethernet LAN with a simple cable. The access point functions as a bridge between the cell and the wired LAN. Stations in the cell and in other linked cells can access all wired LAN facilities. A wireless bridge can be mounted back-to-back with an access point which allows connectivity between a group of networks, linking buildings that are miles apart.

Overlapping Cells

When an area in the building is within reception range of more than one access point, the cells' coverage is said to overlap. Each wireless station will automatically establish the best possible connection with one of the access points. An overlapping coverage area is important in a wireless LAN system, since it allows seamless roaming between the overlapping cells.

Linked cells allow a notebook computer user, for example, to walk from Cell A to overlapping Cell B without interrupting a work session. The change from cell to cell is not noticeable to the user.

When several access points are positioned in such a way that their coverage areas converge, this creates a multi-cell. One way of creating a multi-cell is to connect several access points to external directional antennas instead of the built-in omni-directional antennas. The access points are positioned at different locations in the coverage area with their directional antennas pointing toward the area that will form the focus of the multi-cell. Stations inside the multi-cell area automatically select the best access point to communicate with. This is used in areas where heavy network traffic exists.

A multi-cell has a 1 of access points, so it can provide constant system back-up capability for reliable operation of the wireless LAN.

Roaming

When users with portable stations such as notebooks, notepads and pen-based computers move freely between overlapping cells and continuously maintain their network connection, this is called roaming. Roaming should be seamless, so a session can be maintained when moving from cell to cell. The user may experience a momentary break in the data flow, depending on the traffic. A station implements its roaming capabilities by selecting the access point in its area that provides the clearest signal.

Interference Considerations

Interface includes multipath propagation and sources such as microwave ovens and instrumental, scientific and medical equipment. Microwave ovens can be a source of radio interference since they emit radiations in the 2.4 GHz frequency range. It is necessary to separate the wireless units as far away as possible from microwave ovens. Other equipment using the 2.4 GHz frequency that may cause interference includes instrumental, scientific and medical equipment transmissions and medium-distance radio telephone broadcasts.

Multipath propagation occurs when radio wave signals are propagated in an indoor environment and bounce off reflective and semi-reflective surfaces such as walls, partitions, furniture and equipment. The reflected signals can reach the receiver from all directions and varying strengths and time dispersions depending on the path they travel. Multipath propagation leads to a fading of the transmitted signal. To reduce the problems caused by signal fading, several diversity techniques are used.

Frequency diversity can be achieved by using a Frequency Hopping Spread Spectrum technique. Frequency hopping radios operate well in the presence of interference. The 2.4 GHz frequency band can be divided into 82 one MHz channels or hops.

Radio transmissions are spread over the 79 useable hops, but at any point in time, only a 1 MHz signal is broadcast on one of the hops. The hops are changed 50 times a second in an order specified by the hopping sequence. Consecutive hops are not closer than 6 MHz apart. All stations in the same cell use the same hopping sequence and synchronize their hop timing. If interference is present, it will usually affect only a few hops. Since the hopping sequences are designed so that successive hops are several MHz apart, the interference may interrupt data transmission on a particular hop, but there is only a small chance that it will affect the next hop in the sequence.

Access point diversity is achieved with a multi-cell architecture and its overlapping area coverage. Stations inside the coverage area automatically switch to the access point that provides the best reception.

Antenna diversity is provided by using several omni-directional antennas. These antennas use space diversity to receive signals from different paths. The modem selects which antenna received the better quality signal on a per-frame basis.

Chapter 7
Computer Control

The ultimate control systems employ computer control. A computer can be programmed for a variety of automated control schemes: simple or complex, open loop or closed loop, feedback or feedforward control. A major advantage of computer control lies in the flexibility of programming. The control pattern can be altered by reprogramming the computer.

A few years ago few HVAC systems could afford to dedicate a computer to control automation. Now, microcomputers are widely marketed and much control software has been developed.

Most HVAC applications do not need heavy-duty computing abilities with super-fast calculations and large memory banks. A small microcomputer will handle most tasks. Many HVAC tasks can be easily done with a small dedicated computer system.

Programming

A programming language refers to the way you tell the computer what you want it to do. At the first level up from the actual microprocessor chip is a type of language called machine language. These are commands that instruct the microprocessor to do very basic operations.

Different microprocessors use different forms of machine language called instruction sets. They cannot directly talk to each other. Machine language is made up of binary numbers which contain only 1s and 0s. Each binary word or byte acts like a string of switches and operates on the microprocessors internal digital logic. If a given switch is open, that binary digit or bit is 0. If the switch is closed, the bit is a 1. These different combinations of 1s and 0s mean different things to the processor as depicted in its instruction set.

Translator programs allow the computer to understand a higher more English-like language. BASIC is one of the most popular languages for control applications. It is readily available and convenient to program. BASIC is an easy language to learn and work with. English-like commands are used.

Input Signals

In most HVAC automation systems the computer needs to sense various conditions. Many types of on-off sensors can be connected directly to an input port on the computer. Almost any device that performs a switching function can be used in this way. The computer can tell if each switch connected to its input ports is open or closed.

Several switching-type sensors can be connected to a single input port. Each switching device is one bit of the incoming binary word. For most microcomputers each port can handle one byte (eight bits). Thus, eight switching sensors could be fed into a single input port, and the computer will be able to individually distinguish between them. For example, suppose the following devices are connected to input port A:

- Bit 0 hot water valve,
- Bit 1 window fan 1,
- Bit 2 window fan 2,
- Bit 3 window fan 3,
- Bit 4 air conditioner,
- Bit 5 humidity sensor,
- Bit 6 air temperature sensor,
- Bit 7 water temperature sensor.

The computer is programmed to periodically check the value at input port A. If it finds a value of 0000 0000, it assumes no action is needed and moves on to other programming. Any other value at input port A indicates an out of tolerance condition (one or more switches is closed). The computer is programmed to take an appropriate action depending on the specific value detected.

It is also possible for the computer to be programmed to respond only when a specific combination of input switch sensors are activated. For example, an input value of **0110 0001** might trigger a special action that is not activated by any other input value.

D/A Conversion

A computer can recognize and send out only digital signals for simple on-off functions. This is sufficient for some applications but not others. A continuous range of values is needed in many HVAC applications.

A computer cannot input or output a continuous range analog signal. Conversion between analog and digital signals is needed. Digital-to-analog (D/A) conversion is the simpler of the two processes.

The individual bits of a byte do not have to be treated like independent entities. Each bit has a value dependent on its position within the byte. Each bit's value is also a power of two. For example:

Bit	Value
0	1
1	2
2	4
3	8
4	16
5	32
6	64
7	128

Bit 0 is the right-most bit and bit 7 is the left-most. Then the bit values become

7-6-5-4-3-2-1-0

Bit 7 is worth much more and said to have greater weight than any of the other bits. Bit 0 is worth the least and is said to have the least weight.

To find the total value of a binary word, you add together the place value of all bits that are 1s and ignore the 0s. For example,

0101 1111 = 0 + 64 + 0 + 16 + 8 + 4 + 2 + 1 = 95

In other words, 0101 1111 is 95 times as great as 0000 0001.

Now, assume each bit switches between 0 and +1 v. Then, 0000 0001 would become an analog output of + 1 V. But, 0101 1001 is + 4 V, along with many other combinations, since all of the bits have equal weight.

Some way to weight each bit is needed. One solution is to place a resistor with an appropriate value in series with each bit. Let the lowest resistance value be R. Then the resistances will have the following values.

R, 2R, 4R, 8R, 16R, 32R, 64R, 128R

Now let each resistance be connected to a switch that individually selects +5V or 0V.

This is a simple D/A converter that is made with weighting resistances. The most significant bit (MSB), which is the largest value, is connected to the smallest resistance. The other bits are connected to greater resistances that are multiples of two of the minimum (MSB) resistance value. The least significant bit (LSB) is the smallest value and connected to the largest resistance. This circuit works in accordance with Ohm's law.

If the current is kept constant, the amount of voltage drop across each resistor will be weighted proportional to its binary value as described:

Binary input	Decimal value	Output voltage
0000	0	0 + 0 + 0 + 0 = 0 V
0001	1	0 + 0 + 0 + 0.5 = 0.5 V
0010	2	0 + 0 + 1 + 0 = 1.0 V
0011	3	0 + 0 + 1 + 0.5 = 1.5 V
0100	4	0 + 2 + 0 + 0 = 2.0 V
0101	5	0 + 2 + 0 + 0.5 = 2.5 V
0110	6	0 + 2 + 1 + 0 = 3.0
0111	7	0 + 2 + 1 + 0.5 = 3.5 V
1000	8	4 + 0 + 0 + 0 = 4.0 V
1001	9	4 + 0 + 0 + 0.5 = 4.5 V
1010	10	4 + 0 + 1 + 0 = 5.0 V
1011	11	4 + 0 + 1 + 0.5 = 5.5 V
1100	12	4 + 2 + 0 + 0 = 6.0 V
1101	13	4 + 2 + 0 + 0.5 = 6.5 V
1110	14	4 + 2 + 1 + 0 = 7.0 V
1111	15	4 + 2 + 1 + 0.5 = 7.5 V

The output increases in 0.5-V (LSB) steps. The analog output is directly proportional to the digital input value. To reduce loading problems, a buffer amplifier stage is added to the output.

This system is simple, but with more than a few bits, the resistors have to cover a large range and non-standard values are needed. The resistors have to be closely matched since wide tolerances in the values can affect the accurate weighting of the bits.

A more convenient type of D/A converter is the R-2R ladder network. Here, only two resistance values are needed: R and 2R. This combination allows each incoming bit to be properly weighted without special matching and it can be extended to accept as many bits as needed.

Practical R-2R ladder D/A converter circuits also have a buffer amplifier to prevent loading problems. This is shown in Figure 7-1.

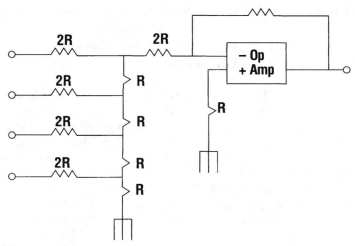

Figure 7-1 Practical R-2R D/A converter

A/D Conversion

In many practical HVAC applications such as closed-loop control, it is necessary to have the computer to examine and interpret analog input data. The analog input signal must be converted to digital form to be usable by the computer. This is done with an analog-to-digital (A/D) converter, which is the opposite of the D/A converter. A/D conversion is a little more complicated than D/A conversion. There are several approaches.

In most A/D converters, the input signal is sampled at a regular rate, usually several hundred or thousand times per second. Each sample is then converted into a proportional digital value. The conversion is not perfect, since digital values change in a stepwise fashion, instead of a smooth linear transition. Intermediate values must be rounded off.

For example, if the A/D converter is set at 1 V/bit, the recognized values will be:

0000	0 V
0001	1 V
0010	2 V
0011	3 V
0100	4 V
0101	5 V
0110	6 V
0111	7 V
1000	8 V
1001	9 V
1010	10 V
1011	11 V
1100	12 V
1101	13 V
1110	14 V
1111	15 V

If the sampled value is 8.5 V, the computer will have to interpret it as either 1000 (8.0 V) or 1001 (9.0 V). If a greater resolution is needed, it is necessary to increase the number of bits which reduces the value of each bit. If a bit value of 0.20 V/bit is used, the values become

0000	0V
0001	0.2 V
0010	0.4 V
0011	0.6 V
0100	0.8 V
0100	1.0 V
0110	1.2 V
0111	1.4 V
1000	1.6 V
1001	1.8 V
1010	2.0 V
1011	2.2 V
1100	2.4 V
1101	2.6 V
1110	2.8 V
1111	3.0 V

Notice how the range is reduced to 3 V for the same resolution. If eight bits are used instead of four, there will be 256 steps instead of the 16 for the four-bit system. If the bit value is 1 V/bit, the eight-bit range will go from 0 to 255 V. If a 0.20 V step value is used, the range will run from 0 to 51 V. Increasing the number of bits increases the cost and complexity of the A/D converter circuitry.

Ramp Type A/D Converters

One of the most common types of A/D converter circuits is the single-slope circuit. Here the analog input signal is fed to an op amp that is wired as a comparator. The input signal is compared to the output of the D/A converter, which forms a feedback loop.

The comparator puts out one bit, which is a one if and only if the input is greater than the output from the D/A converter. This bit is inverted and fed to an AND gate, along with the clock signal. There are four possible input combinations to this gate. Each input combination and its resulting output is as follows:

Comparator	Clock	Gate output
0	0	1
0	1	0
1	0	0
1	1	0

Notice that the gate output is a 1 if and only if both the clock and the inverted comparator output are at logic 0. When the converter output is a 1, the analog input signal is greater than the D/A output.

As long as the inverted comparator output is high, the clock signal is ignored. When the inverted comparator signal is low, the clock signal passes through the gate. The pulses are counted by the binary counter. The output of the counter is converted into analog form by the D/A converter. This analog value is fed back to the comparator.

The counting continues until the D/A output exceeds the external input. The gate is now cut off, so no further clock pulses get through to be counted. The counter now holds a digital value that is proportional to the analog input voltage. This binary number is fed out to the computer input.

The single-slope A/D converter performs the A/D function, but its accuracy can be improved. Better accuracy can be achieved with a dual-slope A/D converter illustrated in Figure 7-2. The basis of this method is to measure the time to charge a capacitor to an unknown voltage and then to discharge to a known voltage.

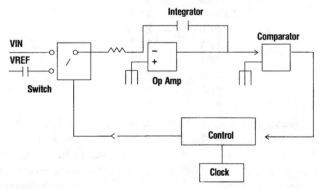

Figure 7-2 Dual slope converter

Due to the several operations involved, it takes some time for each sample value to be calculated. When faster conversions are required, a flash converter is used. This is simply a number of comparators connected in parallel. Each comparator contributes one weighted bit.

Personal Computers

Applications for the personal computer and software packages had their initial appearance in office and business automation. Next, came applications in engineering analysis, production and quality control, and computer-aided design (CAD). Another major application area is measurement and control. Much of this evolved from the extension of PCs to the laboratory, collecting data and monitoring the status of laboratory processes and equipment.

The introduction of industrial software and PC versions allowed the user to incorporate microcomputers directly in control applications or indirectly through programmable or process controllers.

The digital interface of PCs buses requires that analog data acquisition use an analog-to-digital converter to transform measurement data into a PC bus-compatible format. Once the data has been acquired, the information can be easily transformed, analyzed, stored, or displayed. Complex algorithms can be updated to the control functions in real-time or close to real-time.

A major application of PCs in industrial plants is in configuring control systems. Another large application is data acquisition, either as a front end to programmable logic controllers (PLCs) or process controllers or in direct connection to a process through input/output (I/O) boards.

Originally PCs were used primarily to monitor and control simple processes. Today, they are used along with and in place of other major types of monitoring and control devices, such as PLCs or process controllers.

Industrial PCs are available in several different configurations; desktop, portable-battery powered, rack mounted and National Electrical Manufacturers Association (NEMA) enclosured. A typical configuration for control devices in U.S. plants is rack mounting. NEMA enclosures are generally required by plants with corrosive environments.

Most applications software developers have aimed their products at the IBM-PC family because of its market share and its availability in industrial versions capable of surviving the characteristics of plant environments. These computers are provided with cooling removable filters and special cases.

An important development in industrial PC growth was Intel's first 32-bit microprocessor, the 80386. Because of its greater computing power and speed, the 80286 allowed close to real-time as well as multitasking capabilities.

PC Control Evolution

In the late 1970s several companies began marketing PCs, including Apple, Tandy, and Commodore. During this time the use of PCs in data acquisition and industrial control was not an important consideration. Apple started a trend for applications in data acquisition when it began incorporating extra slots into the chassis for adding extra memory or peripheral devices.

The IBM-PC was introduced in 1981 and it also included expansion slots that could be used for analog interfaces. Shortly after the introduction of the IBM-PC, several companies offered add-in accessory boards including analog input/output and IEE-488 interfaces. Most of these initial offerings supported Apple personal computers. Today many companies provide analog interfaces for several different PCs; most of these are designed for the IBM-PC family.

Automated data acquisition systems and personal computer-based data acquisition has received support from many companies including Hewlett-Packard, Fluke, and Burr-Brown. Most vendors have adopted the IBM-PC format for their analog interfaces, primarily due to the acceptance of the IBM-PC family in scientific and engineering applications.

There is also a growing ease in the implementation of PC-based data acquisition systems. As new data acquisition hardware and software products appear, they feature higher performance, lower cost, and ease of use. There have been many improvements in speed, number of channels, resolution, ruggedness, and processing intelligence. Many of these systems tend to compete with specialized test, measurement, and control systems.

Declining costs and more powerful software have resulted in new applications for PCs, as replacements for oscilloscopes, data loggers, and chart recorders. However, these standalone units can have a negative image because of their inflexibility.

A major use of PCs has been in development work, including the development of control applications for other devices as well as performing calculations in support of engineering. Personal computers are becoming part of a movement to build fully integrated data handling systems.

PCs are often used as an operator interface for PLC applications. Here, they are used to change setpoints, start and stop motors, and control valves. The PC with the proper software can be used to replace control and graphic panels as well as annunciators.

PCs have been used to control furnaces in petrochemical plants. The flexibility of the PC is important in many of these applications. It has been particularly useful in those processes which are subject to frequent major adjustments.

PCs have been used in tandem with PLCs for automatic crane control in the manufacture of films and for kiln control in electronics manufacture. This combines the ruggedness and quick scan time of the PLC along with the data storage, manipulation, and reporting capabilities of the PC. PCs are particularly useful in their ability to perform precalculations on data.

In the chemicals industry, PCs are used to reconstruct gas chromatograms from data supplied by remote instruments. PCs have also been used as digitizers for pressure waves in explosives testing.

Some industrial PCs employ a cooling system where the air is drawn in through a filter and returned out through the disk openings. The opposite path is used in office machines.

PCs in Integration

In many areas there is a push for complete integration. The PC has the ability to collect data and automate functions inexpensively in a network environment. A PC network can provide the data sharing that is basic to integrated systems.

A wide array of software development/cost difficulties are possible. One side of this spectrum is improved flexibility and the ability to optimize performance for each application.

Languages and Programming Techniques

A critical part of software development hinges on the issue of languages. There is a trend toward standardizing one or two programming languages. Established standards are important and the most popular languages are BASIC or FORTRAN. Once useful code is written and running in a particular language, there is a tendency for more code to be written in that same language. FORTRAN and BASIC are not the fastest, or the most powerful, or the most elegant languages. They have been in use for some time, and much control code has been written in these languages.

Many are familiar with them and feel comfortable writing programs in them. BASIC, in its many versions, is by far the most widely used language. It meets the needs of many applications, but it does have problems.

The most critical problem with BASIC for control or monitoring applications is its speed even in the compiled version (usually BASIC is interpreted) and even when the BASIC programs can call assembly language subroutines. The slow execution is compensated by fast I/O calls.

The BASIC language is a good all around choice due to the number of drivers and compatible I/O systems available for it which far exceeds similar offerings for other languages.

Although the Instrumentation Society of America extensions of FORTRAN have given the language real-time capabilities, it still has the problem of strong hardware dependence. Programs written in FORTRAN for use on one kind of machine generally must be rewritten if they are to run on another type of machine.

Among the newer languages, C, especially in a UNIX or Xenix environment, and Pascal offer considerable improvements in programmer productivity and in their ability, in some versions, to handle real-time multitasking applications.

Packaged Software

Packaged software continues to grow in variety and sophistication. They provide relief from the high cost of in-house software development.

Database packages are widely used for everything from parts lists and maintenance inventory control to storage and the manipulation of local process data. Graphics packages for the PC are also widely used. Their ability to display data in a readily usable format and to assist in such tasks as the production of flow diagrams continues to grow.

Statistical packages can be used to analyze trends and failure modes. Project management and scheduling packages are useful for planning and managing HVAC modifications and system retrofits. Custom programming may be needed in some cases, but there are many ready-made packages that are easily configurable to specific applications.

There is a trend towards more high-level application languages for ladder logic or flowcharting for control applications which uses a CAD-like programming environment. These are nonprocedural languages as opposed to languages such as BASIC or Pascal which list the procedure to be followed. Ladder logic has been in use longer than digital computers, but the newer generation of nonprocedural languages only requires the user to list the desired result rather than the desired procedure. Used in conjunction with artificial intelligence tools, the nonprocedural approaches to machine instruction promise to greatly reduce the time required in writing customized software.

Programmable Logic Controllers

A programmable logic controller (PLC) is an industrially ruggedized computer-logic unit that performs discrete or continuous control functions. These units were originally intended as relay replacements in the automotive industry. PLCs can now be found in some part of almost every type of industry.

The programmable controller is sold as standalone equipment by at least 10 major control equipment manufacturers. Several other companies produce PLCs for original equipment manufacture (OEM) applications.

The programmable logic controller is sometimes referred to as a PC, for programmable control rather than a PLC. Since the term PC is also used to designate personal computers, use of it in this way can cause confusion and should be avoided.

Programmable controllers were originally used by General Motors in 1968 to save the costs of scrapping assembly-line relays during model changeovers. Table 7-1 lists some of the highlights in the evolution of PLCs. The automotive industry adapted the PLC because of the massive changes that had to be done for a model change.

Besides cost reductions, programmable logic controllers offer other benefits over traditional relay logic. They are relatively easy to program and reprogram after installation. Table 7-2 compares the types of available control logic for HVAC.

Table 7-1 Evolution of Programmable Logic Controllers (PLCs)

Year	Description
1968	PLCs developed for General Motors Corporation
1969	First PLCs manufactured for automotive industry
1971	First application of PLCs outside the automotive industry
1973	Introduction of smart PLCs for arithmetic and other operations
1975	Introduction of analog PID (proportional, integral, derivative) control
1976	First use of PLCs in integrated manufacturing systems
1977	Introduction of small microprocessor PLCs
1978	PLCs widely accepted, with sales near $80 million
1979	Integration of plant operations using PLCs
1980	Introduction of intelligent input and output modules
1981	Data highways used to connect PLCs
1982	Larger PLCs with up to 8192 I/O points

Table 7-2 Control Logic Types

	Relays	**Solid-State Controls**	**PCs**	**Mini-computer/ Workstations**	**PLCs**
Hardware cost	Low	Low-Medium	Low	Medium	Low to Medium
Versatility	Low	Low	Yes	Yes	Yes
Reusable	No	No	Yes	Yes	Yes
Space required	Largest	Large	Medium	Medium	Smallest

The programming language that is used in PLCs is based on familiar relay wiring symbols. PLC reliability is good and there is minimal maintenance involved. They have the ability to communicate with other computer systems. The unit cost is moderate to low with a rugged, modular package design.

The first- and second-generation PLCs were built through the 1970s and 1980s. There are several sizes to select from. The smallest programmable controllers are primarily a relay replacement unit with few additional functions. This is a basic bare-bones, inexpensive controller.

Besides relay replacement functions, the controller may include counting, timing, and complex mathematical functions. Most mid-sized PLCs can perform PID feedforward along with other control functions. Larger PLCs have data highway capabilities.

PLC Components

The functional parts of a PLC include the power supply, I/O, central processor, memory, and programming and peripheral interfaces. (Figure 7-3) The power supply may be integral or separately mounted. It provides isolation to protect the solid-state components from voltage spikes. The power supply converts the power line voltage to those required by the different internal circuits.

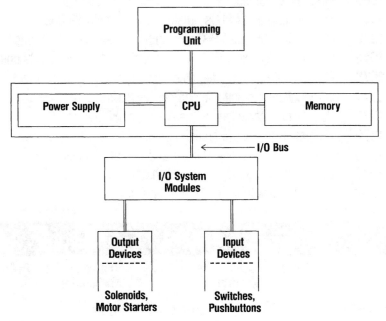

Figure 7-3 PLC Structure

PLCs have high-ambient-temperature specifications. This is an important difference between programmable controllers and personal computers.

As the I/O is expanded, some PLCs may require additional power supplies. This supply may be separate or part of the I/O structure.

Inputs

These are real-world signals and can be analog or digital. They are presented to the programmable controller as a varying voltage, current, or resistance. Analog signals include those from thermocouples (TCs) and resistance temperature detectors (RTDs). Variable frequency signals are accepted along with those digital signals from pushbuttons, limit switches, or relay contacts.

An additional type of input signal is the register input. This reflects the computer nature of the programmable controller. The register input is a parallel digital input. It accepts a collection of digital signals delivered to the PLC at the same time. This could be a binary coded word or number that is compatible with the register input port.

198

Outputs

The three categories of outputs are discrete, register, and analog. The discrete outputs can be pilot lights, solenoid valves, or annunciator windows. Register outputs can be used to drive panel meters or displays. The analog outputs can supply signals to variable speed drives or to I/P (current to air) converters for controlling valves.

The I/O systems are modular, so a system can be arranged for multiples of I/O bits. These modules can be composed of 1,4,8, or 16 bits and plug into the existing bus structure. The bus structure is a high-speed multiplexer that carries information back and forth between the I/O modules and the central processor unit.

One function of the I/O is the ability to isolate real-world signals, 0 to 120 V AC, 0 to 24 V DC, 4 to 20 mA and 0 to 10 V from the low level signal levels of 0 to 5 V DC in the I/O bus. This is done with optical isolators.

Central Processor Unit

The central processor unit (CPU) performs tasks such as scanning, bus traffic control, program execution, peripheral and external device communications, data handling and execution and self-diagnostics.

Central processing units have used TTL (transistor-transistor logic), ferrite cores, CMOS (complementary metal oxide silicon) logic or microprocessor-based (VLSI). The microprocessor-based systems are more powerful and more flexible.

The scan time of the PLC indicates the time it takes for the programmable controller to interrogate the input devices, execute the application program and provide updated signals to the output devices. Scan times vary from 0.1 milliseconds per 1K (1024) words of logic to more than 50 milliseconds per 1K of logic.

Although scan times are a performance measure, there are other factors involved. The word size varies from 4 to 32 bits depending on model and manufacturer. Special features, such as full floating point mathematics, have different processing times and may generate longer scan times.

Memory Unit

The memory unit of the PLC acts as the library where the application program and executive program are stored. The executive program is the operating system for the PLC. It interprets, manages, and executes the application program. The memory unit is also where the input and output data are temporarily stored and used by the CPU.

Memory can be volatile or nonvolatile. Volatile memory is erased if power is removed. Most units with volatile memory provide battery backup to prevent the loss of data and control in the event of a power outage. Nonvolatile memory does not change state on a loss of power and is used where the extended loss of power may be possible.

The basic programmable controller memory element is a word or a collection of 4,8,16, or 32 bits. As the word length increases, more information can be moved or stored.

Most programmable controllers provide the equivalent of 32K of 8-bit memory locations and can execute application programs that are fairly complex with 50 to 100 discrete I/O points.

Programmer Units

The programmer unit provides the interface between the PLC and the user for program development, start-up, and troubleshooting. The instructions to be performed during each scan or cycle time are coded and inserted into memory through the programmer.

Programmers vary from small hand-held units the size of a large calculator to desktop CRT-based units. Many PLCs can also use a personal computer as the programming tool. The programming software is loaded into the personal computer. The actual PLC programming is done over a serial interface in the programmable controller.

A hand-held programmer allows the operator to enter a program one contact at a time. These units are rugged, portable and easy to operate.

The CRT programmer provides a visual picture of the program in the PLC. Ladder diagrams appear on the screen and menu-driven software is used. The screen size varies from 4 to 9-inch (100 to 225 mm).

The CRT screen typically shows 8 rungs of ladder logic by 11 contacts across. The ladder diagrams can be placed into the real-time mode, which allows visual contact status. Some CRT programmers provide complete documentation capability, including ladder diagrams, cross-reference listing, and I/O listing.

Since PLC languages are designed to emulate the relay ladder diagrams, this format is simpler to understand and maintain compared to computer programming. The program can be easily modified or tuned to improve the control program performance.

A major difference between PCs and PLCs is the sequential operation of the PLC. The program operations are performed by the PLC in the order they were programmed. (Figure 7-4) This allows straightforward programming and execution with indexing for real-time control applications.

Self-diagnostics aid in the troubleshooting and repair of problems. Most PLC components are modular and easy to isolate. These remove-and-replace system modules include diagnostic techniques.

CRT programmers are portable but weigh 45 to 60-pounds (20 to 27 kg). A modem connection allows these CRTs to be used at remote locations for programming and troubleshooting.

Memory storage capabilities include floppy disks or magnetic tape. With floppy disks, programs can be copied from one disk to another and verified without the need for loading the PLC's memory. Another useful feature is automatic documentation of the program. This is done by a printer attached to the programmer.

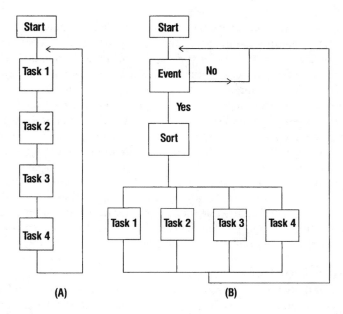

Figure 7-4 PLC sequential execution (A) and PC multiple loop program flow (B)

Off-line programming allows the user to write a control program on the programming unit. Then, the unit can be taken to the installed PLC and load the new program. Online programming allows tuning or modification of the program while the PLC is controlling the process.

CRT programmer software is available that operates on a personal computer. The program behaves like a CRT programmer, but may not have all of the functions that the CRT programmer does.

Peripheral Devices

Peripheral devices include programming and operational aids, I/O enhancements and computer interface devices. Programming aids provide the documentation and program recording capabilities. Some devices can program different manufacturers' PLCs, but most are dedicated to a single supplier and specific models. Programming aids include the PC-compatible software that allows the programmable controller to be emulated by a personal computer.

Operational aids include resources such as color CRTs and support programs that allow the operator access to internal parameters. The operator can read and modify timer, counter, and loop parameters.

I/O enhancements include several types of modules for intelligent or remote I/O capabilities. This includes I/O simulators which can be used to develop and debug programs.

The computer interface devices allow peer-to-peer communications with one programmable controller connected directly to another, as well as network connections.

Programmable controllers have been extended to include mathematical functions, file manipulations, analog and high-speed signals. Programmable controllers are linked together in networks and connected to computer systems and specialized interface devices. These include displays, high-speed control cards, and interfaces to specialized sensors such as bar code readers and RF identifiers.

These added capabilities allow the application of programmable controllers in many areas outside of traditional assembly type service. They are used for continuous process control, high-speed packaging, energy distribution, and automated warehousing. Many of these applications strain the capabilities of ladder logic for program design and implementation.

The software available for programmable controllers has extended ladder logic to serve these other applications. Ladder logic advances include indirect addressing, program flow modification, mathematical/scientific instructions and communications with intelligent devices. The ladder logic segments can also be organized using sequential flowchart languages.

Graphic Control

One of the ways of representing program requirements is through the use of state diagrams and data flow diagrams. These are structured documents that use symbols for terminators, data transformations, data storage and triggers for control flows. Table 7-3 includes a list of these symbols which are called primitives.

Table 7-3 Some Primitives

Primitive	Description
Control Flow	Data triggering
Data Flow	Represents data movement
Data Store	Indicates how data is archived
Input/Output Data Transform	Computes outputs from outputs
State	User-identifiable system mode
Terminator	Defines items outside the system
Transition Action/State Transition	Causes a state change while actions occur

Control flows describe when actions occur. Data flows and data transformations describe how the system operates.

Extensions to Ladder Logic

Relay ladder logic depends on commands which energize, latch, and unlatch coils. Other ladder logic commands are used to read inputs, outputs and internal bits.

Timers and counters require ladder logic commands for the timing and counting of signals. In mid- and large-size controllers, these ladder-logic-style commands are supplemented with assembly-type instructions. Instead of working on bits, they operate on complete words. Some ladder logic extensions are similar in character to higher level languages such as FORTRAN, PASCAL, or APL.

A more primitive language is assembly language. This type of symbolic programming language can be directly translated into machine language instructions which operate the CPU. Assembly language is machine dependent. Table 7-4 shows some typical assembler-style instructions used in ladder logic extensions.

Table 7-4 Typical Assembly Language Instructions

Arithmetic	add	Adds two registers
	mul	Multiplies two registers
	div	Divides one register into another
	ave	Averages contiguous registers
Logical	and	Sets a bit if two bits are true
	or	Sets a bit if either of two bits is true
	not	Reverses the state of a bit
Registers	mov	Moves a bit from one register into another
	bcd* convert	Converts a register from bcd to integer
Queues	Fifo Load	Adds an element to a waiting line
	Fifo Unload	Removes an element from a waiting line

*bcd = binary-coded decimal

Indirect Addressing

Larger systems require a large number of rungs or lines in the ladder diagram. The larger this number is, the more likely there is that one of them could contain an error. Indirect addressing is a way of simplifying and reducing the complexity of large programs. If a HVAC system has similar 20 loops and the code for one loop is about 200 rungs, the result is a 4,000 rung program.

If one rung needs to be changed in each loop with indirect addressing, there is only one rung to change. Without it, modifying the control would require modifications in each of the 20 loops. Indirect addressing provides a means for using fewer instructions than physical coding would use.

In the ladder diagram, the operation of indirect addressing can be done with a loop counter and jump instruction in the ladder rung. When the loop counter trips, the jump instruction runs jumping to a subroutine. This subroutine can be the same for all of the loops which simplifies the programming.

Indirect addressing requires more processor time than physical addressing. Direct addressing is also simpler to conceive.

Program Flow

Some ladder logic extensions modify the program flow. These include subroutine instructions such as GOSUB or JSR and GO TO commands. Conventional relay logic commands execute sequentially. There are no changes in program flow since each instruction is executed in every program scan.

The use of a GO TO command means that some parts of the program will not be executed during the scan. The program flow will jump over the commands between the GO TO and the destination of the GO TO command.

Other devices that serve the same purpose are the jump and GOSUB commands. A subroutine can be executed by a JSR (jump to subroutine) or GOSUB. When this occurs, the program flow goes to the line indicated in the command and executes the instructions that follow.

Modular Programming

Programs with many subroutines are called modular programs. A subroutine allows the same code to be run for different purposes in different places of the program without recopying the code. Fewer lines of code mean less time to make modifications and maintain the code. The subroutine can be run conditionally. This reduces the scan time. The details of the program are in the modules which simplifies the programming task.

A program is broken up into separate subroutines or modules. Common loop logic can be written as a single subroutine. Another subroutine could be reading the temperature or humidity signals. A third subroutine might be the use of the logic outputs to communicate with other equipment. An executive routine can be used to call each individual subroutine when that subroutine needs to be called.

Other assembly language commands can perform data manipulation, including arithmetic and Boolean operations. A register can be assigned for a counting application.

Several commands are used to perform queue management. These include the LIFO and FIFO commands which load and unload the queue. A queue in a programmable controller is usually a series of words which are loaded into registers to form an array. The queue management command will insert or extract a word from the array.

Communication with Devices

Some devices that need to communicate with the programmable controller mimic a set of racks. The racks are then scanned. This provides a simple interface to these devices, but it decreases the number of racks that the programmable controller may connect to.

These devices physically plug into the controller like an I/O card and communicate using some type of serial communication. One common type of card is able to communicate using RS422 or RS232.

The user can program these cards with ladder logic and jumpers on the card to allow communication with any device using this type of serial communication.

Counter cards count pulses which can occur too quickly for the programmable controller. The counter cards allows a block of words to be read or written to the card. A block transfer statement is used to initiate the read or write.

Other methods to read signals that occur faster than the scan time of the programmable controller include time-scheduled interrupts, programmable interrupts and latching hardware.

The programmable interrupt is an immediate update input instruction. When this instruction is encountered, the controller stops scanning the program and checks the state of the interrupting input. Latching cards are useful when the time between pulses is longer than the program scan time.

Another updating method is to use another programmable controller. A smaller, quicker controller may be enough to detect short pulses. It is programmed to act like a counter or a latching card.

Graphic Languages

There are several types of graphic, flowchart-like languages. They allow easy maintenance of the program for the language is designed for sequences of operations. Ladder logic is based on relays which operate in a nonsequential manner. The sequence of control is usually tracked by a counter or by latching cards. The order of the sequence is difficult to quickly determine from the program.

The international standard for a graphic programming language is the International Electrotechnical Commission, Standard 848. This language has several basic symbol blocks or primitives. These are steps, transitions, directed links, branching sequences and simultaneous sequences.

A step is a small program. The program can be written in any language. A transition initiates a move from one step to another. When a step is connected to a transition, the steps can occur sequentially. This is known as a directed link. Steps can also branch, depending on some other condition which allows an alternate sequence to occur. It is possible to program concurrent or simultaneous steps. This allows several concurrent steps to be executed.

Soft Logic

ISaGRAF is a typical Soft Logic package. The workbench section provides PLC programming and is Windows 3.1, 95 or NT PC compatible. It includes graphical editors for programming sequential function charts, function block diagrams and ladder diagrams. There are also text editors for writing the instruction list, structured text and C-code. Other editors are available for variable declarations, analog conversion tables and I/O configuration.

Additional tools include an online debugger, simulation, cross reference, project management, document generation, backup/restore and graphic debugging animation. This methodology will catch most syntactic errors. Hardware independent code is produced and both compiled or interpreted code can be generated.

System Design

The methodology starts with the preparation of a detailed project specification. Distributed applications can be edited, simulated or debugged on the same Windows screen.

The project manager module allows the application specification to be divided into smaller functional modules. The user defines each of these modules, their operations and their interaction to form the complete application.

All variables are declared or imported in a dictionary. Then, during programming, a mouse click inserts the variable in the program. Any external database can be imported to build the dictionary.

The next step is the actual programming of the various functional modules. This can be done using one of the supported languages.

IEC 1131-3

IEC 1131-3 is a worldwide standard for the industrial control programming interface. Modular techniques and the declaration of variables are used to structure each program. IEC 1131-3 also structures the way a control system is configured.

In 1993 the IEC issued the IEC 1131-3 standard which is a specification of five PLC programming languages that can be freely mixed to define automation and control procedures. The IEC 1131-3 languages are:

- Sequential Function Chart (SFC),
- Function Block Diagram (FBD),
- Ladder Diagram (LD),
- Structured Text (ST) and
- Instruction List (IL).

The Sequential Function Chart (SFC) divides the process cycle into a number of well-defined steps, which are separated by transitions. SFC is the core language of the IEC 1131-3 standard. The other languages are used to describe the actions performed within the steps and the logical conditions for the transitions.

The Function Block Diagram (FBD) is a graphical language which allows the user to build complex procedures with function blocks from a library, wiring them together on screen. The ISaGRAF library has more than 60 blocks ready to use and users can enlarge the library by writing functions and function blocks in LD, FBD, ST, IL PLC languages or C.

The Ladder Diagram (LD) is one of the most familiar methods of representing logical equations and simple actions, contacts representing input arguments and coils representing output results.

Structured Text Editor

Structured Text (ST) is a high-level structured language with a syntax similar to Pascal, but it is more intuitive to automation applications. This language is used to implement complex procedures that cannot be easily expressed with graphical languages. It uses IF, THEN, ELSE, FOR and WHILE statements.

Functions and function blocks written in C can be called directly from any of the five supported languages. C functions and function blocks can be used to access any of the resources of the target system.

System tasks that are written in C can be used to display graphics, interface to networks or disks, or perform algorithms for control. These tasks can be started from the application using C function blocks. C function blocks can also send or receive any data from user tasks.

A library manager allows the programmer to create a library of functions and function blocks written in C. The user can also create a library of functions and function blocks in FBD, LD, ST, IL languages.

An I/O connection editor provides a link between hardware independent logical variables and the physical I/O channels available on the PLC/PC. The programmer does not need a knowledge of high-level software. The I/O variables are specified in the IEC 1131-3 standard.

PLCopen is a vendor- and product-independent worldwide association supporting IEC 1131-3. By implementing this standard on many program development environments (known as Program Support Environments or PSEs in the IEC 1131-3 terminology), users can move between different brands and types of control with little training and easily exchange applications.

An open automation control system with EC 1131-3 requires less support and has greater throughout. PC-based control introduces cost savings by integrating the operator interface and control software on one PC and in one database. This can result in development time savings of 50%. Self-documentation can include project descriptor, project architecture, modification history, I/O wiring lists, dictionaries and cross references. Graphic printouts include sequential function charts, function block diagrams and ladder diagrams.

Besides verifying the syntax of the source code entered for any of the supported languages, a test can be made of the application, verifying that the various functional modules are properly linked to each other. When an error occurs during verification, an error message appears and links you to the error in the source code. The generated program can be either downloaded and run on the target hardware or simulated with the workbench. The compiler uses an optimizer for the optimization of binary decision diagrams.

Simulation

Without a target hardware platform, the programmer can validate the application with the simulator. This allows structural and functional tests for each module.

The simulator makes it easier to trace the program execution and view the status of internal variables. The I/O hardware can be simulated and internal status and variables manually forced.

During simulation editors can be opened in the debug mode to see how programs are executed including SFC active steps and LD coil values. Passing from off-line mode (editing) between online mode (simulation/debug) is automatic.

Application level debugging allows you to test the software and hardware. The I/O configuration can be tested separately from the application execution with the reading of inputs and forced outputs.

Each module of the application can be debugged separately and each I/O element can be integrated in the execution independently and online. The online modifications can be performed on-the-fly. Breakpoints and cycle-to-cycle modes can be set/reset online.

The hardware link between the debugger and the target can be an RS-232 serial link, Ethernet link or a network such as CanBus, FIP, LonWorks, NetBios or Profibus.

The target system configuration can be an industrial computer, micro controller, or PLC equipped with analog and digital I/O modules running MS-DOS or a multitasking operating system such as Windows NT, OS-9, VxWorks. Hardware supporting an ANSI C compiler can also be used. The target operating system can emulate Windows 3.1, 95 or Windows NT.

Communication and I/O

The link between the PC workstation and the target can be achieved in several ways. A simple RS-232 serial connection (MODBUS) can be used which is similar to the way PLC terminals are connected to PLCs. Other links include Ethernet TCP/IP, Canbus, FIP, LonWorks and Profibus. Network addresses are defined for each variable or group of variables.

The target system can have any type of plug-in I/O, remote I/O and field-buses. This includes Bitbus, Canbus, FIP, InterBus, LonWorks, NetBios and Profibus. Any other fieldbus can be integrated in the software. I/O can be digital, analog or character strings. Working with a supervisory and control package, communication with a supervisor can be done using a MODBUS link.

PLC Installation

PLC installation is similar to the installation of relay or other control systems. Safety rules and practices include correct grounding techniques, disconnect devices, selection of wire sizes and fusing. PLCs can often be re-retrofitted into existing relay enclosures.

Disconnect switches and master control relays should be hardwired to cut off power to the output supply of the PLC. This is necessary because most PLCs use triacs for their output switching devices and triacs are likely to fail on or off.

PCs and PLCs

PLCs are a type of microcomputer but there are several differences between PCs and PLCs. The PLC has always been designed for real-time operations.

Most PLCs have internal clocks and watchdog timers built to control their functional operations. PCs evolved from basic computer tasks such as word processing and accounting.

The CPU in a PLC is programmed to scan the I/O for status, make sequential control decisions, implement those decisions and repeat the procedure within the scan time.

PLCs are designed to operate near manufacturing equipment so they can function in hot, humid, dirty, noisy, and dusty industrial environments. Typical PLCs can operate in the temperature range of 32 to 140°F (0 to 60°C) with relative humidities from 0% to 95% noncondensing. The electrical noise immunities are comparable with those required in military specifications and the mean time between failures (MTBF) ranges from 20,000 to 50,000 hours.

Control Evolution

Today's control logic is rapidly moving from PLC to PC-based or smart control. The first evolution was from relay panels to PLCs. PLC architecture has its limits since it was invented to replace large relay panels. Flexibility and simple programming were the main advantages. Relay panel logic was programmed by hardwiring. PLCs could be programmed by software. PLCs offered a new way to control applications as well as providing improvements in logic solving.

A relay panel is a parallel processing system. For each output, the logic path consists of easy-to-identify dedicated inputs. To calculate the system reaction time of a specific control module, the relay delay times in the control logic path are added together. Each output has a different reaction time based on how many relays are in the control logic path.

PLCs changed the mechanization for simple process control dramatically because they use a single microcontroller to calculate all the outputs based on the input status of the system. The controller became a sequential single process system that reads all the input states first, then performs the calculation, and finally updates all outputs. This is a 3-step process and is repeated as fast as possible.

The time a system needs to execute one cycle is the PLC scan time and is influenced by several factors such as the microcontroller speed and the amount of code to be processed. The PLC scan time must be faster than the slowest reaction time requirement for an output in the system.

Optimized Logic

The early PLC microcontrollers were not very powerful and memory was expensive. The use of the logic controller was limited to logic control. There was no intent to exchange data or information with other devices. Not enough calculation and memory resources were available.

The design was based on cycle-time speed. Microcontroller clock frequencies were very slow, and several clock cycles were used to execute one controller instruction.

The hardware architecture was optimized to move data efficiently from the

I/O ports to the logic-solve unit and back. This hardware architecture is specialized and inflexible.

The architecture was optimized for speed and minimal random-access memory usage. This allows most of the microcontroller's time and memory to be used to solve an application's logic.

Tools to program, debug, and manipulate inputs and outputs are necessary to successfully apply and use a PLC. Since the hardware and software structure of a PLC is very specialized, the tools used are unique and limited in functionality.

Several different families of PLCs were developed to handle different requirements in discrete control such as closed-loop processes and motion controllers to control a machine.

Since a PLC has a limited functionality, and its hardware and software are restricted, the only way to communicate to other control function blocks is through hardware interfaces. A PLC is a specialized, hardware-dedicated logic control solution.

PLCs Versus DCS

During the past decade, the capabilities of programmable logic controllers (PLCs) and distributed control systems (DCS) have evolved to the extent that many applications can be handled by both. PLCs were developed as a replacement for relay logic and interlock applications, while DCS systems were developed for process control.

DCS systems are strong in PID-type analog control interfaces. If most of the input/output (I/O) control is digital or discrete, a PLC is often used, while if the I/O is mostly analog, a DCS system is a better fit. This rule is not universal and there are other decisive factors in system selection.

PLCs are likely to be more rugged and may handle discrete logic faster than DCS systems. PLCs generally use languages based on ladder logic and these languages are not well suited to analog process control. Some of these limitations of PLCs are overcome by coupling them to personal computers (PCs).

In the areas of communication redundancy and data security, DCS systems are superior. DCS systems are also superior in the areas of programming libraries, advanced or optimizing control, self-tuning algorithms and in their total plant architecture and information management capabilities.

Smart Control

Resource management requires trend analysis and diagnostic information. Communication throughput and flexibility between different system modules is often needed.

There has been a dramatic development of PC capabilities, together with the low price for hardware and software. The widespread base of this universal microcontroller system points to the use of this architecture as a logic controller. (Figure 7-5)

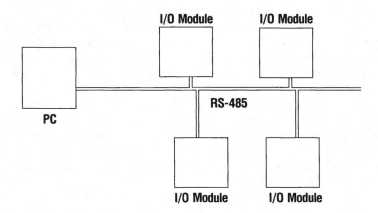

Figure 7-5 PC and remote I/O modules

One concern was the reliability of PC systems. The hardware platforms were originally designed for an office environment, and the mass storage units had a relatively low mean time between failures. The operating systems for open microcontroller systems were not as stable as needed and were not designed for repeatable deterministic behavior.

Over a period of time, several different open hardware platforms that met different requirements were developed. Today, there are PC-based microcontroller systems of different sizes and different environmental specifications. These products include small embedded controllers, powerful industrial PCs and even units that can be used in a space shuttle.

Operating systems include Windows NT, QNX, and IRMX. Because of the almost unlimited resources of the microcontroller in today's PC-based controller platforms, the need for specialized dedicated controller hardware is limited. There are usually enough resources available to execute layers and other services that help modularize a system and support the use of inexpensive, high-volume commercial software modules.

Control Software

A basic piece of control software would do exactly the same job as a PLC. It would be pure logic control. It reads inputs, calculates outputs, and updates outputs.

Additional software modules would be designed to communicate efficiently with other software modules over standard software interfaces. Program development tools and aids would also be included. Additional functional modules can be attached for increasing the information throughput and the flexibility of the control system.

The idea of running different system modules on the same hardware platform has become a viable solution. Today it is common for real-time logic, configuration management, operator interface, system development, and debugging tool modules to be executed on the same hardware platform.

Additional software modules include standard software application packages such as spreadsheet applications or databases. Module functionality varies from system to system, and in some cases they can generate important system management information. In other cases, they track the process for trend analysis and event logging.

The open architecture of PC-based control allows users to choose from a large base of commercial available software. The commercially available operating systems usually have built-in network capabilities that allow the control system to be integrated with a higher-level system. By networking several hardware platforms together, redundancy or distribution of the control functions can be achieved without any additional cost.

The different modules need to be able to work in relationship to each other. They also have to support an information system that includes four major categories of information: real-time process, diagnostic, configuration, and information data. Each category must be treated according to its speed requirements and data size so the overall system performance is achieved. This means there must be an overall concept of how the control solution is modularized in different modules and how different information types are handled.

Conventional PLC architecture is not designed to handle this amount of data and does not include an information feature. All of the data is treated the same way in a PLC, which is not appropriate for an efficient control solution.

Smart control used with advanced networks defines the next generation of control solutions. The wheel of technology spins faster today. This means the adoption of new concepts will be much faster than when they moved from relay panels to PLCs.

PC-based Control for HVAC

Some manufacturers are modifying their control practices by upgrading heating, ventilation, and air conditioning systems with PC-based control. In one project managed by Buchanan Enterprises, the goal was to achieve highly integrated plantwide controls with the effective monitoring of energy used for heating and ventilation.

The HVAC system had seven primary fan rooms, outbound bays, and other process units. The primary fan rooms needed their older pneumatic controls replaced. Data acquisition and control was hardwired from each field device to the input/output devices and to the PC in each fan room substation.

The project used NT in a PC-based control system with DeviceNet communications. Additional benefits included lower installed costs with less wiring and a true distributed intelligence. Communications were improved as well as the diagnostics between devices. Incorporating all control, operator interfaces and data acquisition in a single system allows a leaner overall operation. Hardware savings were almost 25%.

Paradym-31 was used for PC-based control. It includes ladder logic, sequential function charts, and function-block diagrams. It can also incorporate user-defined function blocks, which allow users to customize the application with specific software tool sets. Using PC-based control with the EC-1131 standard and NT as a shop-floor platform can significantly decrease integration costs for plant-level planning and diagnostics.

PC-based Control Solutions

PC-based control has matured into a viable control solution. It can provide control capabilities traditionally missing in more common methods. PC-based control is being used in applications ranging from material handling to discrete assembly.

Some users find that pure PC-based control brings real benefits, while others find blending traditional forms of control like programmable logic controllers with PCs is the best solution.

Mixing and blending control system types is common. Many products complement a complete PC-based control solution and these products must work together to be effective.

Engine Flexibility

A PC-based control engine runs on a commercial operating system and it can operate with many different programs. Operating systems such as Windows NT have hundreds of off-the-shelf application programs. A PC-based control engine used on unmodified Windows NT can use any of those programs to customize the control to the specific application. Engines built on Windows with real-time extensions or on a real-time system may be less open to integration with off-the-shelf programs.

PC-based control engines offer increased options in programming software. Most systems today utilize languages such as Visual Basic, Visual C++, C, or ladder logic. There are also a number of methods to simplify program control including sequential function charts, function-block diagrams, structure text and flowcharts. A system may use a combination of several languages and program control methods. The programming language should fit the application's function.

Applications programmed in different languages can reside simultaneously on the same system. In PC-based control, the information sharing between programs has been simplified. ActiveX controls allow small, specialized software components to be created that can be reused to develop many customized applications.

I/O independence is another feature of most PC-based engines. It allows users to choose the I/O and network solution best suited for the application rather than using a proprietary solution. This has been facilitated by open, multisupplier network technologies such as DeviceNet, ControlNet, and Ethernet.

Open Systems

Today, there are many manufacturers of computer-based systems for commercial and industrial control. Applications include boiler and chiller controls, heating and air conditioning and fire alarm systems. In all of these control and monitoring systems, a common element is some means of communication to link the individual components into a coordinated system.

Most control components are capable of connection to some type of communication network, either proprietary or open. Control system elements need to pass information among themselves to perform their assigned tasks. One example is an outside temperature sensor for hot water reset, which signals a boiler controller to produce hot water at the correct temperature to satisfy a building's heating load.

The diverse base of installed equipment from multiple vendors can be expensive to interface. The installation of control equipment from another manufacturer requires new skills.

Also, the product life cycle for these systems is becoming shorter. As parts and service for older systems become too expensive, it is often more economical to install new systems that can provide better performance and easier maintenance at lower cost.

Once a proprietary control system is installed, it must be expanded or modified with proprietary hardware from the original manufacturer. The desire to interface computer control systems from different manufacturers which allows expansion and modification with other vendors' equipment has resulted in the movement to develop open systems for commercial and industrial control.

The open system concept allows the connection of control system components from a different manufacturer to operate as a single, integrated system in a manner transparent to the user. The system components can then be purchased competitively.

The open systems concept has been developed for IBM personal computer systems. Since these computers are based on open specifications, any manufacturer can build systems around them. There is now a great impetus toward system interoperability in a market that was once dominated by a few large manufacturers' proprietary systems.

The concept of open systems depends on the Open System Interconnection (OSI) reference model for computer communication standards. The OSI model is a seven-layer communication protocol architecture.

Each of the layers has its hierarchical function with defined interfaces to higher and lower layers. Not all layers are for a functional communication standard. When several of the layers are used in conjunction with programming object models to define the meaning and format of the data, a high level of multi-vendor interoperability can be achieved.

In an ideal open systems architecture the devices communicate with each other using a standard protocol as a native language, without translation. This ideal system would have interoperability at all levels.

Commercial open systems can be divided into two major types. One arrangement is to interface workstations and control system components from one manufacturer with other manufacturers' components using gateway hardware or software on the network segments. The native language is a proprietary protocol. This is not a true open system.

The proprietary communication protocols and database structures must be made compatible, using translation in the form of the gateways implemented in hardware or software. The devices in a single communication segment must be from the same manufacturer. These devices include terminal units, chillers, fire alarm panels, security access controls, generator controls, fume hoods and programmable logical controllers.

In the other type of open system, the communications between workstations and field panels take place on a common bus using native non-proprietary protocols. The field panels can be setup internally to perform the gateway functions, or they can use external gateways to communicate with other manufacturers' equipment using a proprietary protocol. Protocols provide for the data exchange of a well-defined set of programming objects and services.

One problem is the downloading or modification of field panel algorithms from remote workstations provided by a different vendor. Typically, this programming is done using each manufacturer's proprietary protocol. It cannot be done from another vendor's workstation, even though these components can communicate for other operations. Another problem involves the interface. It may not be able to support all of the protocol functions and services due to cost considerations. A controller may provide monitoring and control functions but not support the alarm and event-handling services.

In Europe, open systems development has resulted in Profibus and FND. In the United States, the Distributed Network Protocol (DNP) and Utility Communications Architecture (UCA) were recently developed for industrial and utility Supervisory Control and Data Acquisition (SCADA) applications.

Commercial data communication protocols for building automation and control networks include BACnet, sponsored by ASHRAE, and LonTalk, sponsored by the Echelon Corp. The BACnet protocol has been adopted as an ANSI/ASHRAE.

Echelon and other fieldbus technologies are already de-facto standards. The growth of this technology is in the hands of sensor/actuator and MMI vendors. From a building automation system integrator point of view, this is the technology for the next 10-20 years.

LonWorks

LonWorks is a sub-system designed by Echelon Corporation. It uses a proprietary chip (the neuron chip) with a unique address and a communications protocol. The chips are manufactured by several companies, such as Motorola and Toshiba.

LonWorks allows a seamless marriage to another system that is designed using the same parameters. Screens with information from more than one system will all appear in the same manner.

LonWorks can be incorporated into a system in different ways. Echelon designed the neuron chip to communicate with field variables, run process control loops and communicate with other neuron chips. This chip was to be used for monitoring and control at various levels of the system. Some manufacturers even use the neuron chip in the main controller of their EMCS system. While these systems use LonWorks, they are not true LonWorks systems.

The neuron chip is programmed with a variant of the C language called neuron C. It can also be programmed using drag and drop software from Dayton General Systems which allows the programming to be done much faster.

HVAC System Renovation

The 14-story Renaissance style Knickerbocker Hotel in Chicago's Gold Coast district decided to improve guest services and reduce operating costs by automating the building's mechanical systems. The new building control system focuses on traditional heating, ventilation, and air conditioning (HVAC) tasks.

A local LonWorks independent developer, TMI, offered the following interoperable LonWorks-based products: ABB adjustable frequency drives, York direct-fire absorption chillers and RCS temperature sensors. Sharing a single, common twisted pair (78 kbits/s) wire backbone among elements, the system devices work as intelligent zone controls. Instead of a mass of individual wires, the functions are handled by the transmission of messages over a network. The plug-and-play device characteristics aided the connection of the HVAC systems. The connection of the components of the control system was done on site at the Knickerbocker.

The greater part of the automation task involved the multi-zone airhandling devices, refrigeration equipment, and systems for providing chilled and hot water to the rooms in the hotel. The volume of air going into the lobby area is controlled by ABB variable frequency drives. The control of dampers and zones is done with LonWorks-based intelligent actuators.

In common areas such as the ballroom, the lobby, and the restaurant, intelligent space sensors are used to transmit network messages back to the actuators that control the zone dampers.

In the airhandling equipment used, the separation of the hot- and cold-deck dampers caused a synchronization problem. The timing facilities of the neuron chips in the actuators was used to provide assurance that both dampers would always operate in the correct positions. The operation of the heating and cooling valves are also under LonWorks control.

The system operates in a Windows-based environment and allows building engineers to monitor control system data, reset fan temperatures, turn equipment on and off and adjust other operating parameters.

Other options of this LonWorks-based system include guest room automation that uses dedicated channels on the CATV backbone already installed in every guest room. The cable becomes the main communication backbone for guest room services such as monitoring the mini-bar, controlling the heating, and providing real-time security with a door locking system.

BACnet

The American Society of Heating Ventilation and Air Conditioning (ASHRAE) completed its 8-year effort to develop the common protocol called BACnet. The goal was to establish interoperability between DCS equipment manufacturers. Echelon is the standard at the sensor bus level.

The complete standard covers about seven different data link layer models. But, a BACnet compliant product on an ARCNET level and a BACnet compliant product on a sensor level does not establish interoperability. Certification of system components as BACnet compliant can also be a problem.

At minimal compliance, it does not make much difference what kind of system or what communications protocols are used in collecting data or in issuing field commands. What is important, is what data is gathered from any given piece of equipment, and how that data is structured. Using this strategy and basic network tools, information gathered from multiple field systems can be presented as a single system, to multiple users.

One problem with a BACnet minimum level of compliance, is that it is not easy to make programming changes. In most cases, programming must be done at the PLC connected to the field equipment.

Full compliance allows programming as well as monitoring and control functions at the user level. Most manufacturers have offered BACnet compatibility at the minimum compliance level.

ControlNet

From proprietary to public standard, the proliferation of industrial networks offers user benefits, but unfortunately many of the benefits are spread across different networks. ControlNet is an automation and control network that tries to combine the benefits of several networks onto one link for increased system performance. It is designed as a control network to reliably predict when data will be delivered and ensure that transmission times are constant and unaffected by the devices connected to the network.

ControlNet is a real-time, control-layer network for the high-speed transport of time-critical I/O and message data. This includes uploading/downloading of programming and configuration data and peer-to-peer messaging, on a single physical media link.

ControlNet operates at 5 Mbits/second. It allows multiple controllers to control I/O on the same wire. This is an advantage over other networks, which allow only one master controller on the wire. ControlNet also allows multicasting of both inputs and peer-to-peer data, which reduces traffic and improves system performance.

Determinism is the ability to reliably predict when data will be delivered. The deterministic performance of ControlNet can be used for both discrete and process applications. Other features include user selectable I/O and controller interlocking update times to match application requirements.

Network access is controlled by a time-slice technique called Concurrent

Time Domain Multiple Access (CTDMA), which regulates a nodes chance to transmit in each network interval. The minimum interval time is 2 ms.

Plantwide network architectures are evolving into three layers. ControlNet fits into the middle automation and control layer for real-time I/O control, interlocking, and messaging. The upper information layer for plantwide data collection and program maintenance can be done with Ethernet. The device layer for the integration of individual devices is handled with the device buses like DeviceNet and Foundation Fieldbus.

ControlNet is useful for systems with multiple PC-based Controllers, PLC-to-PLC and PLC-to-DCS communication. ControlNet allows multiple controllers with their own I/O and shared inputs to talk to each other.

The concept of an industrial control network is to link all devices or stations with one cable. No matter how many devices you are connecting, you just add the physical cable which is needed to connect to the furthest station. Troubleshooting becomes much easier since you can plug and unplug the device from the network. This virtually implies a plug and play ability of the devices.

DeviceNet

DeviceNet is a simple low cost, open industrial control networking system that was first established by Allen-Bradley and now maintained by an independent organization, Open DeviceNet Vendor's Association (ODVA). DeviceNet utilizes the Controller Area Network (CAN) as a backbone and follows the ISO/OSI 7-Layer model stack. The basic DeviceNet characteristics are summarized in Table 7-6.

Table 7-6 DeviceNet Characteristics

3 baud rates — 125k, 250k, 500k bps
24 volts DC, 1% tolerance
Standard 11-bit identifier CAN architecture
64 nodes

DeviceNet is an 8-byte fieldbus system for industrial I/O control such as sensors, switches, bar-code scanners and motor drives. These devices are linked together on a bus using a DeviceNet cable. This cable contains 4 wires: Vcc, CAN-high, CAN-low and Ground. There is a shielding wire around the 4 wires for noise immunity purposes. The cable lengths are given in Table 7-7. Several chip manufacturers such as Phillips, NEC, Intel, Siemens and National Semiconductors supply CAN chips.

Table 7-5 ControlNet
Technical Specifications

Type of Bus	Control I/O data, Programming data on same wire
Bus Topologies	Linear Trunk Tree Star Mix of any of above
Bus Speed	5.0 megabit/sec (maximum)
Length, single segment	1,000 m (coax) @ 5 Mb/s 1,000 m with two nodes 250 m with 48 nodes 3,000 m (fiber) @ 5 Mb/s
Number of Repeaters	5 (maximum) in series 6 segments (5 repeaters) in series 48 segments in parallel
Maximum Length with Repeaters	5,000 m (coax) @ 5 Mb/s 30+ km (fiber)
Types of Repeaters	high voltage ac and dc low voltage dc
Device Power **Connectors (Standard Coax BNC)**	Devices power externally barrel (plug-to-plug) bullet (jack-to-jack) extender (plug-to-jack) isolated bulkhead (jack-to-jack) right angle (plug-to-jack)
Communication Model	Producer/Consumer
Number of Nodes	99 Maximum Addressable Nodes 48 taps (nodes) without a repeater
Size of Data Packet	Variable, 0–510 bytes
Number of I/O Data Points	No limit
Network Update Time (Scan Time)	2–100 msec (user selectable)
Communication Modes (Bus Addressing)	Master/Slave Multi-Master Peer-to-Peer
I/O Data Triggers	Polling Cyclic Change-of-State
Cyclic Redundancy Check	Modified CCITT using 16-bit, polynomial
Application Layer	Object-Oriented: Class/Instance/Attribute Device Object Model using Device Profiles
Physical Media	Coax – R6/U Fiber
Network and System Features	Remove/Insert Devices Under Power Deterministic, Repeatable Intrinsic Safety Option Duplicate Node ID Detection Message Fragmentation (block transfers)

Table 7-7 DeviceNet Cable Length in Meters

Data Rate	Trunk Length		Drop Length	
	100% Thick Cable	100% Thin Cable	Maximum	Cumulative
125K bps	500	500	6	156
250K bps	250	100	6	78
500K bps	100	100	6	39

DeviceNet uses OOT (Object-Oriented Technique) where objects are defined in class, instances and attributes. This allows changes to be made more easily. Four basic objects are used in DeviceNet:

- Identity object – it provides general information about the identity of a device.
- Connection object – this contains the number of logical I/O ports of the device.
- DeviceNet Object – this configures the port attributes such as node address and data rate. It maintains the configuration and status of physical attachments.
- Message Router – this routes received messages to target objects. The priority of access depends on the Message Group the device belongs to.

CAN

The Data Link Layer of DeviceNet is defined by the CAN specification and by the implementation of CAN controller chips. Data is moved on DeviceNet using a data frame. The 0-8 byte size is designed for low-end devices with small amounts of I/O data that must be exchanged frequently. With eight bytes, simple devices can also send diagnostic data.

If two or more nodes try to access the network simultaneously, an arbitration mechanism resolves the potential conflict with no loss of data or bandwidth. In comparison, Ethernet uses collision detection which can result in a loss of data and bandwidth since both nodes have to back-down and resend the data. The winner of an arbitration between two nodes transmitting simultaneously is the one with the lower numbered identifier.

A cyclic redundancy check is used by CAN controllers to detect frame errors. Automatic retries are also used for fault confinement. These methods are mostly transparent to the application. They tend to prevent a faulty node from disturbing the network.

HVAC controls are available for many open systems. The shift from proprietary to open control systems which is similar to the shift that occurred from proprietary mini-computers to generic, IBM-standard personal computers based on open specifications. Another example of an open system is computer-aided design (CAD), where DGS, and DXF file interchange protocols allow the AutoCad and Microstation CAD systems from different software vendors to exchange graphic files.

Building Automation Trends

Open control systems using BACnet and LonTalk are becoming common. Product interoperability depends on the BACnet committee's documents on minimum BACnet performance requirements for different types of equipment and the developers of LonTalk, under the LonMark trade association, developing similar performance requirements for different types of mechanical and electrical systems.

Open system product offerings include building automation systems, DDC controls, security systems and fire alarm systems. Some manufacturers support one or both of the leading commercial protocols, BACnet and LonTalk.

Most building and control system applications can be implemented using the available open system products. Existing computer networks based on commercial data network technology, such as Ethernet and ARCNET, can be used to transfer open systems data. The trend towards using enterprise-wide computer networks for non-traditional applications is growing fast and fits into the open systems concept.

Open System Considerations

Open systems are coming into widespread use and the issues that need evaluation include the following: the protocol functions supported need to include reading digital and analog values, changing setpoints and schedules, and downloading data.

Operator workstation functionality may be limited when communicating with the open protocol equipment. Ideally, conformance with interoperability should be plug and play.

The functionality of the operator workstations are important since system configuration, routine engineering and maintenance functions are not usually performed at a single location.

Several tradeoffs are involved in using open systems. These include performance limitations, system integration, maintenance, and service issues.

The mixing of open system standard protocol capabilities with proprietary system functionality leads to increased costs and creates potential purchasing and implementation problems.

Detailed functions and services must be specified since the standard protocols do not provide the same capabilities. These include monitoring, alarm, setpoint changes and scheduling functions.

The definition of system integration responsibilities is important when using equipment from different manufacturers. These responsibilities include integrating new equipment and the database at the operator workstations to incorporate new requirements. Changes in one manufacturer's hardware or software may affect other portions of the system.

The definition of factory and field testing is also important. This includes the acceptance criteria, based on individual component and integrated system functionality. This functionality must be tested at both levels. It is not enough to test the monitoring and control functions supported by the open systems protocol from the workstation to each field panel. The programming functions required at each field panel location must also be tested.

Maintenance is a critical issue for the successful operation of the system. System maintenance refers to both hardware and software. Hardware maintenance is simplified because of built-in equipment diagnostics and plug-in replacement boards. Software maintenance can be more demanding since there may be multiple databases and application programs to maintain as well as several different programming languages. Electrical and mechanical system maintenance programs can be used for preventive maintenance and trend recordkeeping.

System Hardware

Software is only part of the solution, the hardware on which the software runs is also important. The PC system can be an industrial computer, commercial-grade computer, or open type of controller. The hardware usually depends on the environment in which it will run. Industrial computers are available with shock-mounted hard drives, ball bearing and filter cooling fans, internal temperature sensors, redundant hard drives, system control monitor cards and uninterruptible power sources.

PCs as PLCs

Several companies such as Intellution, Wonderware and SoftPLC have developed programs that run on a PC that emulate a PLC. These programs use ladder logic and allow the PC to function like a PLC.

There is a development platform and a run-time platform, and in some cases an operator interface front end. Most programs such as Paradym-31 from Intellution have a self-documenting capability and can go from ladder logic to other programming techniques. Table 7-8 summarizes the characteristics of these software packages.

Table 7-8 Typical Control Software

	USData Corp	Wonderware	PC Soft International	National Instruments
Company				
Product	FactoryLink, WebClient	FactorySuite 2000	Wizcon for Internet	Lookout
Web interfaces	X	X	X	X
Explorer	X	X	X	X
NetScape	X	X	X	X
Java		X	X	
Visual Basic	X	X	X	
C, C++	X	X	X	X
DDE interface	X	X	X	X
Alarming	X	X	X	X
Diagraming	X	X	X	X
Dial-in access	X	X	X	X
I/O monitoring	X	X		X
Online changes	X	X	X	X
Time trending	X	X	X	X
File manager	X	X		
Report generator	X	X	X	X

Wonderware

Wonderware uses Microsoft Windows for their human-machine interface (HMI) software. Applications include supervisory control and data acquisition (SCADA), production management and resource tracking, machine diagnostics and connectivity to control devices and computing networks. Windows 95 and NT operating systems are used as well as Microsoft BackOffice for industrial automation.

Wonderware software components provide the following system services:

- InTouch is a human-machine interface (HMI) for visualization,
- InControl for Windows NT-based machine and process control,
- Scout is an Internet/Intranet tool for remote data viewing.

InTouch provides a single integrated view of all control and information resources. It allows operators to view and interact with the workings of an entire operation using graphical representations. It includes remote tag referencing, ActiveX support, distributed alarm handling, and distributed historical data.

Online statistical analysis tools help users to achieve better quality, reduce costs and increase yields.

Individual alarms help users to quickly identify which statistical rules were violated. Users can track corrective actions, set limits and special causes on the fly.

A library of 16- and 32-bit I/O servers are used to provide connections to PLCs, RTUs, DCSs, flow controllers, loop controllers and other hardware.

OPC OLE for Process Control is the emerging open standard interface for Windows data acquisition programs. Wonderware's OPC client is called OPCLink. It enables communications with in-process and out-of-process OPC servers.

A high-speed communications protocol, based on TCP/IP, is used to provide performance-optimized data communication in the Windows NT environment for real-time, peer-to-peer communications.

A real-time relational database combines the power and flexibility of a relational database with the speed and compression of a real-time system. Wonderware's process information management system, IndustrialSQL Server provides data access, a powerful relational engine and integration with Microsoft BackOffice.

IndustrialSQL Server acquires and stores plant data at full resolution and integrates real-time and historic plant data with configuration, event, summary and production data. This data can be visualized, analyzed and reported across the enterprise.

InControl is an NT-based real-time open architecture control system that allows the user to design, test and run application programs for controlling a process. Direct interfaces to a variety of I/O devices, motors, sensors, and other equipment are available.

The Internet visualization module acts as a web server add-on and client browser that allows read-only remote viewing of data and visual objects over the Inter/Intranet.

In the client or browser mode, visual objects can be viewed from the server using a standard browser. The software extends a browser's capability by providing a set of user-configurable ActiveX objects such as graphs, charts, and trend elements that can be dynamically linked to the data.

In the server mode a set of web server components runs on Microsoft's Internet Information Server (IIS) and links applications to the Internet.

Chapter 8
Building Automation

Distributed Control

Computer-based systems that need higher performance and greater reliability use a technique known as distributed processing. It has the following characteristics:

- The overall task is divided into smaller segments,
- Each of the tasks is assigned to a different computer, and
- A central master computer or controller ties all the elements together.

Each of the computers can concentrate on its own task. Since each computer is working on only part of the problem and they are all working at the same time, overall performance of the system is improved.

Increased system reliability is achieved with software that allows continued operation if one of the computers goes down. The others continue with all of the tasks with some reduction of performance.

PLCs have not typically been designed to support distributed processing, but distributed systems are being built using technologies developed for PCs. These technologies include faster CPUs and better peripherals, communications, and networking capabilities.

Distributed Techniques

There are several hardware approaches for implementing a distributed system. A standalone controller based on PC technology can be used. It could be a fully functional PC running under a smaller operating system such as ROM-DOS. This would not require a keyboard or a monitor. Software is developed on a host using a standard language such as C. The programs are downloaded over a standard communication line, such as RS-232 or RS-485, and stored in local, nonvolatile memory. When power is applied, this small computer/controller autoboots and executes the stored code. This type of small controller can be used as a node in a distributed processing system. As a node, it would be responsible for the control of the sensors and control elements attached directly to it.

The central controller operates in a system-management mode. It checks on the operation of the different nodes, collecting and processing informational and status data, and updating the operating programs as necessary. This reduces the computing load on the central controller and the performance of the system is improved.

Since a complete computer system controls each node, if the central controller fails, the nodes can continue to operate in a standalone node until the central controller is repaired. (Figure 8-1)

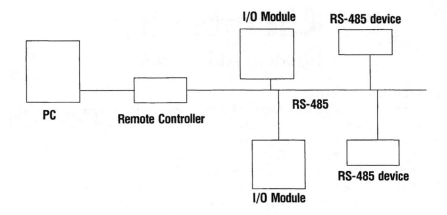

Figure 8-1 A Remote Controller at Each Node Allows the Nodes to Continue to Operate in a Standalone Mode.

If one of the nodes fails, then only the control by that node is affected. The other nodes continue to operate. The nodes can also be used as protocol converters. (Figure 8-2)

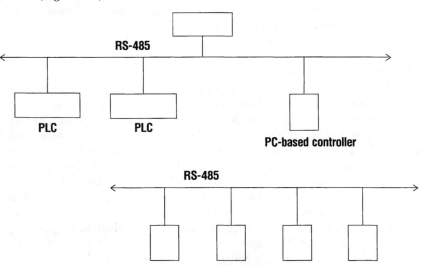

Figure 8-2 Nodes can be Used as Protocol Converters.

Sensor Location

The output of many sensing devices is low and when it becomes necessary to locate these sensors in a remote location there are a number of factors to be considered. Most of these factors are related to the levels of the signals that the sensors produce and include noise, loss of fidelity, calibration and environmental problems. Placing amplifiers as close as possible to the sensor or providing a sensor/amplifier package has been used to address these problems.

An application program has three major components. There is the operator interface which is heavy with graphics, diagrams, and other visual aids. These elements are designed to update the operator and it is often the largest part of the overall program.

Another major portion is the reporting section. This part formats and formulates the outputs. The last component of the program is the code that does the actual work of interfacing with the sensors and applying the control. This part is usually the smallest and may be less than 5% to 10% of the overall code. This last part is the only portion that needs to be downloaded to the remote nodes, so the nodal memory can be quite small.

Another hardware approach combines the computer/controller with signal-processing modules in a single unit. This approach is normally used in situations where the number of inputs and outputs is higher. A single digital or analog input or output is often called a point.

This approach requires more memory which is needed to support the higher point count. These types of systems will often support higher communications speeds in order to keep with the higher point count.

PCs can provide direct digital control and they offer the lowest hardware cost for implementing programmable control. The power of the PC and its availability at modest prices are its main advantages. Today's PCs far exceed the capabilities of the general-purpose computers which were available in the early days of digital control.

Real-time Systems

The term real-time must be defined. A system that responds to a keyboard is not necessarily real-time. A better definition of real-time is a system that guarantees a definite maximum time between an interrupt and the execution of the first instruction of the code designed to respond to the interrupt. The cause of the interrupt may be an external event such as a valve or switch closure or by internal event such as the lapse of a clock-time setting.

PCs use a bus structure with their input/output (I/O) devices attached to this bus. Channel logic is required for each I/O device so it can control the bus.

A co-processor can be used for real-time operation or the PC can function with very short I/O times. Time-sliced buses can also be used.

The PC operating system is a major concern. DOS was designed as a personal operating system and needs a master interrupt control (MIC) to function in real-time. The current focus on PC operating systems has been on more advanced human user functions such as found in the Windows program.

Control Interfaces

Measurement data and control can be found in four classes of equipment. These are sensors, remote terminal units, programmable logic controllers and distributed control systems. Sensor I/O can be found in plug-in boards for PCs.

Some of their important characteristics include common mode isolation, system grounding, and power supply requirements.

In systems where the PC cannot be placed close to the sensors, the use of a remote terminal unit (RTU) may be needed. The RTU has its own enclosure so it can be stronger physically and electrically than the PC. The communications lines between the PC and the RTU must operate in real-time since most RTUs have limited logic and little storage capability. Microprocessing can be added to the RTUs to emulate the performance of programmable logic controllers (PLCs).

Real-time communications capability with the PC is almost always required with the PLC. Some PLCs have analog-to-digital conversion capability with enough logic for at least simple control loops. Table 8-1 compares PLC with PC-based control.

Table 8-1 PLCs Versus PC Control

PLC	PC control
Specialized hardware	Optimized hardware
Inflexible architecture	Flexible architecture
Data exchange using hardware interfaces	Data exchange using software interfaces
Does not share hardware platform	Shares hardware platform with other applications

For proportional control or proportional plus integral control, configuration is done with a PC, host computer, or special terminal.

PID Control

The PID features found in the control loops of today's Direct Digital Control (DDC) systems allow much greater accuracy in control systems compared to that available only a few years ago. When setting up PID (Proportional + Integral + Derivative) loop control, achieving proper operation involves the understanding of the setup parameters and the sequence of implementing them.

Proper operating control can be defined as the ability to control a variable at a given setpoint within an acceptable degree of accuracy. Because of the dynamics of the control system, abrupt changes in setpoint or system loading can cause the controls to oscillate or control with excessive error between setpoint and actual control point. To better understand this process, consider the following terms commonly used in PID control.

Oscillation and Proportional Band

The period of the loop oscillation is the time from peak to peak. All control loops have a tendency to oscillate because of built-in timing constants of the control system components and dynamically changing variables such as setpoint shifts or load changes.

Typical period values found in HVAC control system loops are in the range of 30-seconds to 20-minutes. All loops can be forced to oscillate by setting the throttling range too low (loop gain too high). Loop oscillation is undesirable in control systems and can be eliminated by increasing the proportional band of the loop.

The proportional band is also known as the throttling range. It is defined as the amount of change in the controlled variable required to drive the loop output from 0 to 100%. Systems that are subjected to abrupt changes in load or setpoint typically require a wider proportional band to remain stable during these system upsets. Very fast system response times, such as those found in static pressure control, require much wider proportional bands to prevent excessive overshoot, which leads to oscillation.

The gain of the loop is inversely proportional to the throttling range or proportional band. In general, decreasing the throttling range increases the amount of overshoot. The larger the throttling range, the slower the loop will respond.

A common characteristic of proportional control is an error between the setpoint and control point, which is called offset or droop. As the system load and/or proportional band increases, so does the offset from setpoint. At 100% or 0% loop output, the offset will be equal to 50% of the throttling range. For example, with 10 degree throttling and 100% loop output, the actual control point will be offset 5 degrees from setpoint.

Adding Integral Action

A common characteristic of proportional control is an error between the setpoint and the control point. This is called the offset or droop. It is an undesirable characteristic of proportional-only control loops and can be eliminated by adding integral action to the loop.

The integral component of a control loop will continue to increase or decrease the output as long as any offset or drop continues to exist. This action drives the controller in the direction needed to eliminate the error caused by the offset.

The controller accomplishes the correction by determining the amount of error that exists between the actual value of the controlled variable and the value of the loop setpoint. Then, it acts to reset the setpoint by the error amount over a time interval.

Integral action is sometimes defined in repeats per minutes of rest toward the setpoint. The range of adjustment is usually about .01 to 2.0 repeats per minute.

Integral action causes the controller output to move in the direction of the setpoint by an amount equal to the difference between the loop output when the setpoint is equal to the control point and the actual loop output caused by offset.

Suppose a loop is at setpoint when its output is at 50%. If the offset or error causes the loop output to be at 20% (the proportional term is 30%), an integral value of one repeat per minute will change the loop output 30% per minute in a direction to bring the control point back to setpoint.

The loop output change will occur in increments throughout the time period of a minute. The size of the increments will depend on the controller. If a loop has an update time of five seconds, the 30% change resulting from integral action of one repeat per minute will occur in 12 steps (60/5) of approximately 2.5% each (30/12).

Since the system's dynamics can change with each increment of integral action, the loop may reach setpoint before the full 30% integral change is achieved.

Derivative Action

Derivative action adds the effect of the errors differential or rate of change. This means when an error changes by more than a given percentage during a specified time period, a portion of the error is added to the calculated output to boost the output response. Typically, derivative action is effective only if the loop can respond quickly to a surge in the output.

Derivative action observes how fast the actual condition approaches the desired condition and produces a control action based on this rate. This action anticipates the convergence of actual and desired conditions. It counteracts the control signal produced by the proportional and integral terms. The desired result is a reduction in overshoot.

The loop control encountered in HVAC control does not usually require the use of derivative control. Derivative action is more common in process control, which can involve rapid response times and large overshoots.

Setting Up PID Control Loops

Proper setup of PID features in the HVAC control loops involves the following guidelines. Before stabilizing the loop by increasing the throttling range, measure the period of oscillation which is the time (in minutes) from one peak to the next (one complete cycle).

Next, obtain loop stability using proportional control only. This is done by increasing the throttling range until the loop is stable with no oscillation. Then, add an additional 10% as a safety factor. Some HVAC loops, such as mixed air, can require a throttling range of 25 degrees or more to achieve stability.

If stability cannot be achieved in this way, the system installation and design needs to be reviewed since the addition of integral or derivative action will only cause further instability.

When stable loop operation is achieved using proportional only control, increase the throttling range by 20 to 30% before adding integral control.

The following formula can be used to calculate the initial integral value to be used:

1/ [2x(loop period in minutes)]

Then, monitor the loop control to evaluate the response to this integral value. If the response is slow with integral action, increase the integral value slightly. Sometimes, it is necessary to upset the loop to observe the loop response. This can be done by changing the setpoint to simulate a sudden change in load, while observing the time required to reach the new setpoint.

The integral value should not exceed 1.0 for integral, so it is better to start small rather than overshoot this value. Values between 0.1 and 0.5 will usually provide good control.

Most control loops used for HVAC do not require derivative action. An improper derivative value will produce worse control than none at all. Usually proportional and integral control can provide the needed precision. If derivative is used, the following formula can be used to determine the derivative value:

(loop period in minutes)/8

Self-tuning Loops

Some commercial control systems offer an automatic self-tuning loop feature that eliminates the need to time the loop period, calculate the proper integral value and select the correct proportional band.

The basic principles of PID control and of self-tuning PID loops are the same for all DDC control systems. However, specific details and algorithm design may vary from one manufacturer to the next.

PID control represents a significant advancement for HVAC controls. It is a very effective technique for providing precise control. While self-tuning loops can achieve good control and save time, they can cause some new problems in more complex control schemes.

For example, the use of self-tuning loops in HVAC systems that sequence two or more control valves or dampers can cause an overlap that turns on heating and cooling at the same time. This happens when the self-tuning loop uses an excessively wide throttling range, which reduces or eliminates the amount of required separation between heating and cooling devices.

Another caution involves self-tuning a loop to control two, two-position heating or cooling devices in sequence. This requires a specific throttling range value to obtain the desired sequencing results.

When using self-tuning, be sure to monitor the system performance long enough to ensure the entire control system operates properly and functions as a system. Once the self-tuning PID values have stabilized and system operation has been monitored for proper and stable control, the self-tuning feature may be turned off to prevent unexpected system disturbance from changing the PID parameters in the future.

Self-tuning loops tend to ratchet the PID parameters (throttling range, integral, and derivative) upward at a relatively quick rate in an attempt to achieve stable operation. If any of these values overshoot, it generally takes much longer for the algorithm to bring them back to more realistic values.

PC Hardware

PCs are microprocessor-driven, but the control loop sees the microprocessor as transparent, since application programs are usually written in a high-level language. When a PC is being used in dedicated process control, high microprocessor speed is an advantage because the necessary housekeeping functions can be quickly accomplished and the processor can frequently poll the I/O interfaces to see if attention is required.

A major concern has been the performance of disk drives in dirty or electrically noisy environments, and there are alternatives in the form of optical storage and diskless systems. Dust and dirt filters can be used and are readily available.

Communications

PCs usually have RS-232-C, RS-422, token ring, collision detection and other electrical and low-level logic protocols. Some plants have installed fiberoptic rings running at 16 megabytes, while other plants use much slower four wire telephone links.

There is also a focus on hardware rather than systems, with a strong move to provide standard open buses for I/O attachment. Physical specifications such as Eurocard have been highly promoted.

Programming

The C language is very compact and is a powerful tool for a skilled programmer. FORTRAN provides more self-documentation. Basic is not as powerful, but it can be made to be understandable with the addition of remarks. These languages are not specifically oriented to control applications, and special subroutines are usually required for functions such as reading analog input.

The Instrument Society of America (ISA) has defined a set of FORTRAN sensor I/O calls. These are found in the following standards:
- ISA/ISO 61.1 Industrial computer system FORTRAN procedure for executive functions process input/output and bit manipulation executive functions,
- ISA/ISO 61.2 Industrial computer system FORTRAN procedure for file access and control of file count,
- ISA/ISO 61.3 Industrial computer system FORTRAN procedure for management of parallel activities.

Evolving Controls

The instrumentation used to implement automatic process control is still in an evolutionary process. Early control systems used large pneumatic controllers. These were later miniaturized and centralized onto control panels and consoles.

Then, analog electronic instruments were introduced. The first applications of computers resulted in a mix of traditional analog and newer direct digital control (DDC) equipment. This mix of equipment was cumbersome and inflexible because the changing of control configurations required rewiring.

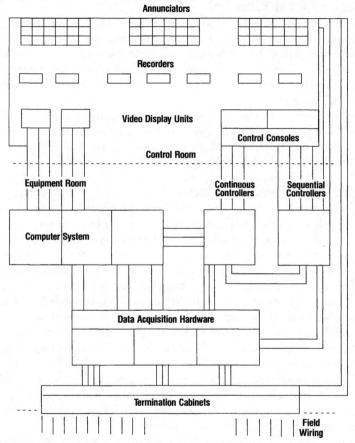

Figure 8-3 Early DCS Where Analog and Digital Instrumentation is Intermixed

Since the early 1970s, distributed control has become an alternate to the other two forms of electronic control equipment used for control; analog controllers and digital computers.

Analog Control

In a large analog control system, there is a great deal of information available to the operator, in a relatively small space. Analog instrument panels offer a limited view of the overall operation. Since the operator should be aware of a number of events occurring at once, particularly when handling an emergency, this type of panel arrangement is often unsatisfactory.

Analog instrument panels also use too much expensive space. The instrument arrangements on the panel are fixed and the panels do not easily permit the relocation of associated devices.

Direct Digital Control

Direct digital control (DDC) and the supervisory control of analog systems by computers are other control options. In direct digital control a computer develops the control signals needed to operate the control devices. In supervisory control, a digital computer generates signals used as the reference or setpoint values for conventional analog controllers. Both of these techniques have been used in the past.

In the early computer applications, cost and delivery times were a problem since each installation was different and required a separate programming effort. The availability of standardized software packages for most of the more routine functions has reduced this problem.

Distributed Process Control

The distributed control system (DCS) is the major form of instrumentation used for industrial control. The equipment in a distributed control system is separated by function and is installed in two different work areas of the installation. The equipment that the operator uses to monitor control conditions and to manipulate the setpoints of the operation is located in a control area or room.

The control conditions can be changed from a keyboard. The controlling part of the system are distributed at various locations throughout the area. They perform two functions at each location. One is the measurement of analog variable and discrete inputs and the other is the generation of output signals to actuators that can change the controlled conditions. The input and output signals can be both analog and discrete.

The information is communicated between the central location and the remotely located controller locations. The communication path may be a cable from each remote location to the central station, or a single cable data highway interfacing all the remote stations.

Since the parts function together as a system, they must be completely integrated and tested as a system. When the components of the system communicate over a shared data highway, no change is required to the wiring when revisions are needed in the control system.

Distributed Control Advantages

Distributed control has several advantages over conventional analog control or digital computer systems. Distributed control can reduce installation costs since less wiring is required when information is transmitted serially across the two wires of a data highway.

Panel space is reduced, and so is the area required to house it. Distributed control hardware costs may be per loop than the equivalent analog control loop but the overall cost including installation is lower.

The operator interface is improved. The group display provides a view of the control loops and their associations. Configuration from a keyboard allows rearranging or adding to the display without the installation of new equipment.

The digital values are quicker to interpret and offer more confidence than a value that must be inferred from the position of a pointer between two scale divisions.

Compared to a digital computer installation, distributed control has cost advantages due to the reduction in wiring. The cost of the computing hardware is also reduced.

In the distributed control system, many small computers are distributed throughout the area installations sharing the work load.

A PC connected into the system can specialize in one or two tasks. It does not have to handle all the information transmittal, data manipulation, and system coordination that a single computer must manage. Even if the central station facilities break down, the control operation will continue without interruption.

The distributed control system is also flexible and relatively easy to expand. Figure 8-4 illustrates a system which could initially consist of only an operator station, a printer, and a control unit. This system can grow to include additional I/O, control units, and operator stations and also interfaces with supervisory computers and PLCs.

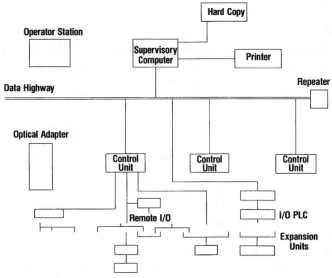

Figure 8-4 DCS System With Control Units, Operator's Station, Printers and Input/Output Units.

The programming effort required for a single computer system is also eliminated. The system programming that directs the operation of the controller electronics, the central station, and the communications links is available as part of a distributed control package.

The programming required to tailor the system can be done without knowing a programming language.

Reliability

Personal computers are more reliable today than when they first appeared. The possibility of failure of a single piece of electronic equipment causing shutdown is greatly reduced by task partitioning.

In the individual electronic units, there is further distribution of responsibility since in each one a number of microprocessors share the functions.

Many suppliers subject samples of their components to time periods of cycling at high and low temperature extremes. The components most likely to fail will usually show up in the first few hours or days of operation. Once past that period, electronic equipment seems to operate almost indefinitely, as long as limits of temperature, shock and vibration are observed.

The advent of large-scale integration (LSI) has improved reliability. As size has been reduced, so has the number of connections and heat generation.

One form of redundancy is controller backup. A backup file is maintained for protection from failures. This allows one set of electronics to be swapped with another set and the system can be up and running as fast as possible. PC equipment is common and inexpensive enough to be supplied on a one-for-one backup basis.

Manual Stations and Analog Controllers

Another type of backup is in the form of manual stations. These are specific to a single control output and allow a local operator to take over from the automatically generated signal at any time. The manual station usually has two dials, one showing variable value and one showing percentage output value. There is a manual adjustment for opening or closing the final control device. Some include a setpoint adjustment, in addition to the final control element adjustment. If a separate power supply is used for the manual station, this type of device can be used to continue operation even if the entire distributed control system fails.

The system can also be designed so that the output signals generated by the distributed control system can be backed up by conventional analog controllers. This allows continuous automatic control, even if there is a microprocessor failure in an output circuit of the distributed controller.

Another technique is to use trip circuits that actuate relays in the event of failure of an electronic controller. The relay contacts can be used to connect actuating devices, either manual or automatic, to permit a local operator to maintain operation.

Reliability and Availability

Availability is as important as reliability and availability, can be defined as the ratio of mean time between failure (MTBF) to mean time between failure plus mean time to repair (MTBF + MTTR). The system will be most available when it is very reliable with a high MTBF and can be quickly repaired, having a low MTTR.

Since a distributed control system is highly modular and contains many printed circuit boards, the time to repair can be short if spare parts are available. Most systems have good diagnostics. The internal failures are reported on the screen, indicating where in the system a failure has been detected and providing some reason to the cause.

The substitution of printed circuit boards can often restore operation. This can be done quickly even in the case of the main computer board. The failed board can then be returned to the supplier for replacement.

DCS Configurations

The interface package serving the interconnections can include PLCs.

Table 8-2 Main Components of DCS

Operator Interface
Keyboard, disc, tape, trend recorder, printer units, communications controller

Computer Interface

PLC, Sensor Interfaces

Dedicated Loop Controllers
For critical loops, card controllers with analog and digital I/O, analog controllers, connected to data highway

Multi-loop Operations Controller
Local control units for other control loops, can also include low level operator's interface

Multiplexer
For non-control monitoring, can also perform PID loop control using the main controller

I/O for the most critical control loops, which should continue functioning even if the central processor or the data highway fails, should be connected to dedicated card controllers. This is the most expensive method of PID implementation in a DCS system, but it is also the most reliable, since a card is dedicated to that loop.

A unit operations controller may be dedicated to a unit operation and all I/O associated with that unit operation is connected to that controller.

One type of I/O that does not serve control functions but serves only monitoring functions is a local multiplexer. This unit does not execute any control functions. It only transmits I/O to and from the central processor. If PID or other control functions are to be performed on this I/O, they must be done in the central processor.

Operator Interface

The operator interface is the area where the operator follows the process, taking advantage of the translation of raw data into trends and patterns. The interface also includes key-entry which allow the operator to enter new setpoints or other parameters. The peripherals include discs, tapes, trend or other recorders, and printer units.

The complexity of a DCS system can vary a great deal. In its simplest form, a DCS system might consist of only a PC display and keyboard and a modular control unit with a point-to-point connection between them.

The distance of the point-to-point connection can be 1000 feet (300 m) or more. There can be from 8 to 32 analog outputs and up to 256 digital inputs and outputs.

Point-to-point wiring can also be used when there are several control units in the system. The remote units may be connected by individual cables to the central computer in a star formation. In this configuration the operator can interface with a number of loops without using a data highway.

Video Display

Most systems use visible cursor control to define screen areas. Some use light pens and others touch-sensitive screens. Most distributed control systems do not accept voice commands or talk back to the operator although the technology exists for both.

Some software gives more emphasis to the keyboard in moving among the display. This is effected by the use of dedicated function keys, addresses based on instrument tag numbers, addressing by group numbers, and function keys that are display-dependent. The latter type relates directly to the display on the video screen. Depressing the key selects the associated screen segment and expands its information content in the next display. Other techniques used for obtaining this type of telescoping effect can include the use of cursor positioning with a mouse, touch-sensitive screens, light pens, pan and zoom type joysticks.

Keyboards

The keys may be movable pushbuttons, or they may be printed squares on a flexible membrane. The membrane type keyboard is similar to those used in microwave ovens and pocket calculators. The membrane-type switch has a flexible hermetically sealed covering. When the pattern of a key printed on the membrane is pressed, a conductive elastomer sheet is pushed through an open-ing under the key, making contact with another conductive sheet under it.

Special function keys may be used by the program. A list of the functions commonly performed is shown in Table 8-3.

Table 8-3 Typical Special Key Functions

Alarm
    Acknowledge, Defeat, Restore
Trending
    Trend time span (Increase, Decrease)
    Range limit adjust (Increase, Decrease)
Mode select
    Automatic, Manual, Cascade, Computer
Setpoint, Output, Bias, Ratio
Setpoint trim-slow and fast entry
    Raise, Lower
Output trim-slow and fast entry
    Raise, Lower
Logic
    On, Raise, Start, Reset
    Off, Lower, Stop, Reverse
Group select
Overview select
Detail select
Channel select

Peripheral Devices

Distributed control systems require bulk memory storage and the device most commonly used is the disk drive. Another bulk memory device sometimes used is the tape cassette.

Printers are a common accessory. They are used to make permanent records of alarm conditions, changes in control parameter values, and trend logs.

Display Formats

Three basic pictorial representations or display formats are used. Group displays show the operating parameters of eight, twelve, or sixteen control loops, usually arranged in rows so that they look like the faces of instruments on an instrument panel. Each of the control loops is represented by bar graphs to indicate the values of the variable and the output signal. A moving index positioned beside the variable bar shows the setpoint value. The actual values for the variable, setpoint, and output percent are shown in or below the rectangular area. Several lines of text allow a tag number and a service description to be shown. The rectangle can change color (usually to red) if an alarm condition occurs.

The array of rectangles resembles a row of instruments on a panel board of analog controllers. From the keyboard the operator can select a loop or an operating mode and choose between automatic, manual, or cascade besides changing setpoint values and output values.

A keyboard procedure called configuration allows any combination of loops to be included in the group display. These can be arranged by association, alarm priority, sequential operation, or other common parameter. A loop may be included in several displays if needed.

The overview display shows only the bare essentials of a number of groups. The setpoint is shown as a straight line reference and the deviation of variable from setpoint appears as a vertical bar. If the group display has eight loops, there will be eight reference lines.

If the variable for a particular loop is greater than the setpoint, a vertical line will rise out from the reference line for that particular loop. If the deviation is in the other direction, a vertical line will drop down from the reference line segment. If the process variable is at the setpoint value, there will be no vertical line.

The overview display shows the condition of all the loops in a number of areas. It takes advantage of pattern recognition and provides a lot of information in a skeleton picture of the present conditions.

Digital conditions may also be displayed on an overview display. Discrete conditions, such as an open or closed switch, can be shown as the presence or absence of a bar rising from the reference line. Sequential events may be displayed as messages that change as the sequence advances.

Detail Displays

The detail display shows a specific loop or control function. A bar graph representation is used with additional information defining constants, limits, and other characteristics of the function. The sources of the signals are also listed on the screen. The configuration of a function is done by moving a cursor and typing in the values on the keyboard.

Other types of display that are used include the graphic display and the trend display. The graphic display allows the actual control flow diagram to be drawn on the screen. Process and control information is included in the diagram, and it can be interactive, changing as real-time information changes. A pipe can become filled with color as a valve is opened. The symbol of the valve can change color, and its condition can be identified with a label that indicates on or off. Graphics are valuable as training or diagnostic tools.

Trend Displays

Trend displays represent a profile of values of a variable showing changes that have taken place over a period of time. Some detail displays provide a real-time trend graph of the variable values during a selectable period. Several trend graphs can be displayed at once, providing a comparison of the history of several variables. Trends over longer periods can be saved on disk memory and displayed on command.

Software Configuration

Configuration involves defining the composition of groups and overviews, trending periods and assigning priorities. The individual control functions must also be programmed.

Data highway systems are configured from the operator station. Files are set up for the various categories of information.

Configuring is usually done off-line and is saved on disk. The items to be configured for a typical system are listed in Table 8-4.

Table 8-4 Configuration Items

Assign station numbers to the highway stations
Define the type of station and priority
Define points to be recorded
Make table of units of measure
Make table of messages
Define tag names
Define displays
Define alarm points and priorities
Define trend time spans and points to be trended
Define sources of inputs and parameters (gain, reset, sensitivity)

Control Configuration

Distributed control systems use a powerful algorithm library. This is called up by answering a number of questions similar to configuring PLCs. What distinguishes the DCS configuration from PLC configuration is the size and quality of the algorithm.

DCS makes it possible to implement more advanced control strategies, including dynamic compensation, lead/lag blocks for feedforward, external feedback for anti-windup and self-tuning. There are also algorithms for analog input, statistical process control, fuzzy logic and model-based optimization.

The CPU operates with a clock to repeat a sequence of programmed subroutines over a time period, typically half a second. This period is divided into a number of segments, or time slots. In each of these, subroutines specific to a control operation are carried out.

The subroutines use algorithms. They are a program of instructions designed to perform a specific function. The time slots may have one algorithm per time slot, or they may have a series of algorithms to accomplish the control action.

Typically, all operations are repeated every half second. At the beginning of the half second period, the inputs are scanned by a multiplexing circuit, and the analog signals are changed to digital values, which are then stored in registers as binary numbers. The transition from analog to digital is performed by an analog-to-digital converter circuit. The signals are linearized, if required, before being stored.

The information about each of the time slots, as well as about each input and output value, is stored in the computer in a database. The functional subroutines (algorithms) are also stored along with the operating program for communications control and information transfer. When a time-slot period begins, the information for that particular time slot is brought from the database into the computer's RAM.

The configuration is checked for the algorithms which are to be used and these subroutines are invoked. Configuration of the controller functions can be built up from a number of basic algorithms such as input modules, alarm limits and error detection. These are sequenced in much the same way as the steps of a programmable pocket calculator.

Another method uses one highly structured algorithm with many smaller subroutines. The first type is similar to the SAMA instrumentation symbology, where the error signal, setpoint, and limit functions are shown separately. The second is similar to ISA symbology, where a circle encloses the controller function.

The first type is built up of small subroutines and is very flexible. The second, since it is more structured, simplifies the configuration.

Algorithm Libraries

Algorithms fall into three distinct control types: analog control, digital control and sequential control. Table 8-5 shows some algorithms in a typical distributed control library.

Table 8-5 Typical Algorithms

PID/ratio/gap	Analog input/output
PID – gap	Digital input/output
Adaptive tuning	Output tracking/limiting
Ramp generation	Feedforward
Sum/divide	Setpoint ramping/ramping
Mass flow	Anti-reset windup
High/low select	AND, OR, exclusive OR, NOT
Dead time	Switch/latch
Comparator	Timer/counter/sequencer
Median	Lead/Lag
Exponential/log	Square root

Some PID algorithms are shown in Table 8-6. Advanced libraries include such activities as sample and hold, decoupling, relative gain calculations, non-linear process adaptations, dynamic adaptations, model-based optimization, fuzzy logic, statistical process control, matrix-based envelope control, and neural networks for artificial intelligence systems.

Table 8-6 Advanced PID Algorithms

Proportional (Gain)	Integral (Reset)	Derivative (Rate)
P/I/D + Adaptive Gain	P + Bias	P + I
P + I + Adaptive Gain	P + D	P + D + Adaptive Gain
P + D + Bias	I + D	P + I + D (PID)
PID + Adaptive Gain	PID + Ratio	
PID with Bumpless Transfer	(Auto Ratio)	

These algorithms are connected together by software addresses to form a combination of functions. This is sometimes called soft wiring. The information resides in memory registers; the software provides pointers toward specific bits or bytes in the register at a specified address. Input addresses can correspond to the output registers of other algorithms. Configuration information is entered from the keyboard, using a fill in the blanks procedure.

Input/Output

Input and output terminations are made at separate terminal boards. Connections are usually made in a separate termination cabinet, filled with rows of terminal strips.

Analog input and output signals will usually be carried on shielded, twisted pairs of copper wire. Digital inputs and outputs, either 120 volt AC or 24 volt DC, can be carried on twisted pairs, which do not have to be shielded. Analog signals should never be run in proximity to alternating current wiring.

Controllers operate almost universally on 1 to 5 volt signals, so the most common input is a 4 to 20 mA current signal, which produces a 1 to 5 volt input across a 250 ohm resistor mounted on the input terminal board. Some distributed control systems accept low-level signals from RTDs and thermocouples, performing signal amplification with their input circuitry.

Signal conditioning can include square root, linearizing signals from thermocouples and resistance thermometers and dampening noisy inputs. These are selected during configuration.

Separate terminal boards can be used for digital input and output signals. Optical isolation is often provided. Here, the input signal causes a light emitting diode (LED) to be energized. A photoelectric device is then energized from the LED. This device actuates a logic input circuit to signal a digital input.

Pneumatic outputs may be developed using a servo-operated 3 to 15 PSI (0.2 to 1.0 bar) analog controller. Other controllers can provide position-type analog damper control, developing the signals for an electric reversing motor-type drive unit.

Communications

The standardization of communication links is essential to DCS. Communications systems differ according to the functions they must perform. Management systems are concerned with the manipulation of large quantities of data in a relatively uncritical time framework.

At the upper levels of management systems for an organization are the two international standards:

- Manufacturing Automation Protocol (MAP),
- Technical Office Protocol (TOP).

The MAP protocol was first defined in 1983 by General Motors to allow an open data exchange between the different automation equipment used by the company for manufacturing. Several manufacturers of automation equipment offer MAP interfaces.

MAP is designed for the transmission of medium to large data quantities. The costs per node are relatively high. It can be run on both wide and carrier band systems and uses the Token Passing method (IEEE 802.2) of bus access.

In applications where time is more critical, an Enhanced Performance Architecture (MAP/EPA) is available. This uses a MINI-MAP 3-layer system that is based on Layers 1, 2, and 7 of the ISO/OSI-Model. It is suitable for applications with transmission times of approximately 25 msec.

TOP

The Technical Office Protocol (TOP) was originally defined by Boeing for technical/scientific office automation. It is designed for the exchange of documentation, including CAD/CAM text and graphics.

A range of transmission media and access methods is supported including

- CSMA/CD on both carrier and wide band,
- Token Passing Ring on twisted pairs and
- Token Passing Bus on wide band.

In Layers 3 to 6 of the OSI Model, MAP is very similar to TOP.

Fieldbuses

Fieldbuses are used to link sensors and actuators with the system intelligence which may consist of DCS systems with PCs and PLCs. The fieldbus allows the sensor information to flow into the factory information system. A fieldbus must handle signals securely and efficiently. Only at the application level is transparency to other networks required.

There have been several efforts to develop a standardized international fieldbus which is open and flexible. The fieldbus must allow the instruments of various manufacturers to operate on the same network. These networks are then linked to production automation and data processing systems. Fieldbuses must fulfill standardized demands for modularity, reliability, interference resistance, ease of installation, maintenance, and transparent operating. Fieldbuses offer a number of advantages over hard-wired installations:

- Bidirectional digital communication,
- Simplified cabling systems,
- Easy expandability and retrofitting,
- Improved safety through self-monitoring and
- Higher resolution of setpoint values.

Digital fieldbus sensors can monitor their own performance and provide diagnostics. They can also be remotely configured and interrogated. The needed intelligence can be integrated directly into the sensor or distributed between it and a controller which also supplies the sensor with electric power.

Factory or manufacturing automation is generally characterized by high transmission rates, short reaction times, and bus lengths up to 300 m. Process control requires an average cycle time of 100 ms with bus lengths to 1500 m. Among the evolving fieldbus standards are the open manufacturer fieldbuses such as the Rackbus, closed fieldbuses, and the MODBUS protocol. More recent entries include FIP-BUS (France), Profibus (Germany) and the MIL-Bus (Great Britain).

Based on a German standard, Profibus is supported by Rosemount/Fisher Controls, Siemens, Endress & Hauser, Yokogawa and some 200 others. Profibus is essentially a token bus, where the traffic director sends a token to one slave device at a time, authorizing it to talk on the network. It is similar to the Manufacturing Automation Protocol (MAP).

Another technique is Fieldbus, which was formed from the alliance of the standards groups InterOperable Systems Project (ISP) and WorldFIP North America, which combined protocols. This is based on the technology of the French standard FIP, supported by Honeywell, Telemechanique and others. This protocol is better on fast loops since it obtains information using interrupts. It enables field devices to talk to each other without going through the host.

Another technique evolved from Allen-Bradley, Numatics, Modicon, Pro-Log and others interested in the integration of discrete devices. This protocol was called Controller Area Network (CAN). CAN is known as a defacto standard under the name DeviceNet.

Profibus

Profibus is a German standard, DIN 19 245 Part 1 (Layers 1 and 2) and Part 2 (Layer 7). Working models were first presented in 1989 and pilot installations were installed in 1991. The first field instrumentation with the Profibus protocol took place in 1992. Profibus is an open system. It was originally standardized in 1989 as DIN 19 245 and in July 1996 as EN 50 170.

Profibus is designed for communication with simple field instrumentation.

Master-slave access is used with central polling and broadcast messages. Object addressing is used with the MAP/EPA three-layer protocol of the physical, data link, and application layers.

The simple communications protocol can be implemented in most of the microprocessors in modern field instruments. Up to 127 devices can be accommodated and up to 32 devices can be designated active token passers.

Profibus uses a hybrid access method combining a master-slave system with decentralized token passing. Active devices are the masters and passive devices perform as slaves. A logical token ring is built up by the active devices. When an active device receives the token, it has the right to transmit. It can communicate with all devices on the bus for a fixed period of time. When this time elapses, the token is passed on to the next active device. Single master operation is possible using a lone master with no logical token ring.

Communication services include up– and downloading of data, create/delete, read, write, reports, notification, acknowledge, after-event-conditions and status.

Profibus Variations
• Profibus-FMS

This is is the universal solution for communications tasks at the upper level (cell level) and the Field Level of the industrial communication hierarchy. It can handle the communications tasks with acyclic or cyclic data transfers at medium speed.

Profibus-FMS is included into the European Fieldbus Standard EN 50170.
• Profibus-DP

This is the performance optimized version of Profibus for time-critical communications between automation systems and distributed peripherals. It can be used as a replacement for the parallel wiring of 24-V and 4 to 20 mA measurement signals. Profibus-DP is included into the European Field Standard EN 50770.
• Profibus-PA

This is designed for connecting automation systems and decentralized field devices. It is based on Profibus-DP (EN 50170) and allows transparent communication along with interoperability and interchangeability of the field devices from different manufacturers.

A GSD file is used to identify a Profibus-DP/PA device as master or slave. This is a simple ASCII text file containing device data such as identification information, transmission speeds supported, data format and time required to respond.

Profibus is a stable protocol with many protocol chips available. It has an installed base of more than 2 million devices.

Software configuration can be done with Allen Bradley's plug and play software or the Siemens COM Profibus package. Most of the field devices use digital technology internally. This eliminates the need for digital to analog converter circuitry.

The Profibus standard defines two variations of the bus cable. Type A is recommended for high transmission speeds (>500 KBaud) and permits doubling of the network distance in comparison to Type B.

A class 1 master can communicate actively only with its configured slaves and is able to communicate in a passive way with a class 2 master. The class 2 master is the supervisory master. They can communicate with other class 1 masters, their slaves and his own slaves for configuration, diagnostic and data/parameter exchange purposes.

No signal ground/reference wire is used in Profibus cables. Isolation of the interface circuit from the local ground must be done with opto couplers. This reduces the possible common mode voltage between transceivers to a minimum.

Profibus is found in many applications in factory, process and building automation. These include automotive manufacturers at General Motors in Europe as well as Breweries such as Bitburger and Guinness. It is also used for building management at the Russian Kremlin building.

FIP-BUS

The Factory Information Protocol (FIP-BUS) resulted from the efforts of a number of French companies. The FIP-BUS is both a protocol and interface. It uses a control scheme based on broadcasting. A bus arbitrator is used and most of the data transmitted on the bus are variables. Each of these variables is identified by a name.

A variable is processed by one transmitter but it can be read by a number of receivers. The use of broadcasting means that it is not necessary to give each device a unique address, but all devices must be capable of receiving the variable intended for them.

The bus arbitrator uses three operating cycles. First, the bus arbitrator names a set of variables. Then, the bus arbitrator calls on request variables from every device. Next, the arbitrator gives the right to transmit to a device.

The control functions are freely distributed among the various devices on the bus, but the data exchange timing and transmission are strictly controlled. This makes FIP act like a real-time distributed database.

The FIP-BUS physical, data link, and application layers conform with the French Standards UTE 46-601-606 and the OSI model. Other standards in the UTE 46 series involve the glass fiber physical layer and network management services. Figure 8-5 shows a typical FIP-BUS.

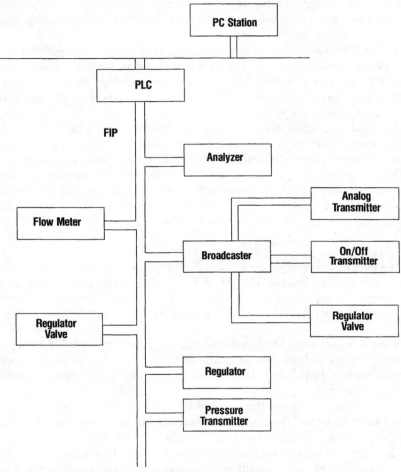

Figure 8-5 Typical FIP-BUS Network Structure

MODBUS

MODBUS is a transmission protocol for control systems. No transmission medium is defined, but RS-232C, RS-422, or 20 mA current loop are all suitable at the transmission rates which the protocol defines.

MODBUS is accessed on the master-slave principle. The protocol allows one master and up to 247 slaves. Only the master can initiate a transaction, which is a query/response where only a single slave is addressed or a broadcast/no-response where all slaves are addressed.

Table 8-7 MODBUS, Profibus, and FIP-BUS Features

Feature	MODBUS	Profibus	FIP-BUS
Initiators	Gould-Modicon Gould, AEG	Profibus User Org. Siemens, Bosch, Klockner-Moller	FIP Club Peugeot, CEGELEC, Telemechanique
Reference or standard	Gould Reference PI-MBUS-300	DIN 19245	UTE 46 (France)
Specifies	Protocol	Interface, protocol and application	Interface, protocol and application
Cable	Not specified	Shielded, twisted pairs, coaxial	Twisted pairs, fiber, coaxial
Length	15m – RS-232C 1200m – RS-422 1000m – current loop	0.2 – 1.2km Maximum 4.8km with 3 repeaters	Maximum 2km
Interface	RS-232C, RS-422 20 MA current loop	RS-485	Proprietary FIP
Transmission rate kbit/s	0.6 to 19.2 depending on distance	9.6, 19.2, 93.75 187, 500, depending on distance	31.25 – 1000, depending on distance
Number of devices	1 master maximum 247 slaves	Maximum of 32 per bus, expandable to 127 devices if repeaters used	Maximum 256
Coding	Configurable, ASCII or RTU	No-return-to-zero	Proprietary FIP
Address range	247	Maximum 32 active and 127 inactive devices	256
Transmission method	Not specified	Half-duplex, asynchronous, no-jitter synchronization	Proprietary FIP
Bus access	Master/slave	Hybrid token ring	Central bus arbitrator
Network management services	Polling diagnostics test reports	Request/send functions	Real-time scanning polling error checking

Some characteristics of the MODBUS protocol are fixed, such as frame format, frame sequences, handling of communications errors, and exception conditions and the functions performed. The characteristics that are user-selectable include the transmission medium, transmission characteristics and transmission mode. The MODBUS protocol provides frames for the transmission of messages between master and slaves. The information in the message includes the address of the intended receiver, instructions for the receiver, data needed to perform the action and techniques for checking errors. Communication errors are detected by character framing, parity checks and redundancy checks.

The functions supported in the MODBUS protocol are designed as control commands for both field instrumentation and actuators. Coil control commands are used for reading and setting a single coil or a group of coils. Input control commands allow the reading of inputs. There are also diagnostics test, report, and polling control functions.

The MODBUS is slower than Profibus and FIP-BUS, but many manufacturers of PLCs produce equipment with the MODBUS protocol. Implementations of the MODBUS protocol are used on low speed point-to-point serial links or twisted pair multidrop networks and it has been adapted for radio, microwave, public switched telephone and infrared applications.

Rackbus

The Rackbus is a serial, character-oriented bus designed to link intelligent transmitters to external networks via gateways. The bus allows the data provided by the measuring points to be available to the supervisory system. The transmitters may be remotely configured. The protocol is open so any vendor may use it.

The Rackbus uses Layers 7, 2, and 1 of the OSI model. Twisted pairs are used as the transmission medium and may be up to 55 yards (50 m) long.

Rackbus operates on a master-slave principle. The master is a special module which provides the link to external networks. Up to 64 transmitters can be used as slaves. This provides a polling potential of 128 measuring points with twin-channel transmitters. Each transmitter has a unique address. There are interfaces for sixteen binary outputs or eight analog outputs which can be current or voltage.

Gateways link Rackbus to other bus systems, while interpreting the protocols and handling requests. The gateway is interfaced to the fieldbus with special modules for RS-232C, RS-422, and 20 mA current loop. The gateways offer several services including device polling. Here the Rackbus transmitters are polled from a scan list.

The gateway allows the Rackbus to be linked into any supervisory network including the following:
- MODBUS gateways with a standard RS-232C port connection to a PC and a MODBUS protocol port for connection to a PLC.
- Profibus gateways with RS-485 interfaces. The gateway becomes a Profibus slave.
- FIP-BUS gateways.

Foundation Fieldbus

The Fieldbus Foundation is a group of suppliers and users dedicated to establishing a single international fieldbus standard. This represents instrument, actuator and other control equipment manufacturers as shown in Table 8-8.

Table 8-8 Fieldbus Foundation Products

Product	Manufacturer
Analyzers	Honeywell, K-Patents, Yamatake-Honeywell
Bridges	Pepper+Fuchs, Yamatake-Honeywell
Chip Sets & Roundboards	Delta, National Instruments', Smar, Yamaha, Yamatake-Honeywell
Configuration Software	National Instruments', Smar, Yamatake-Honeywell
Controllers/Simulators	Fieldbus Inc.
Control Valves	Yamatake-Honeywell
Flowmeters	ABB, Yamatake-Honeywell, Yokogawa
Flow Transmitters	Honeywell, Yamatake-Honeywell
Host Systems	Elsag Bailey, Fisher-Rosemount, Foxboro, Honeywell, National Instruments', Smar, Wonderware, Yamatake-Honeywell, Yokogawa
Input/Output	Intellution, Pepper+Fuchs, Smar
Interfaces	Groupe Scheider, Pepper+Fuchs, Rockwell, Smar, Softing
pH Transmitters	Rosemount Analytical, Yamatake-Honeywell
Pressure Transmitters	Elsag Bailey, Foxboro, Fuji, Honeywell, Rosemount, Smar, Yamatake-Honeywell, Yokogawa
Temperature Transmitters	Elsag Bailey, Honeywell, Smar, Rosemount, Yamatake-Honeywell
Tools	Delta, Fraunhofer
Valve Controllers	Fisher, Valtek
Valve Control Systems	Valtek
Valve Positioners	Foxboro, Honeywell, Masoneilan, Palmstiernas, Smar, Yamatake-Honeywell

The Foundation Fieldbus is a digital, serial, communication system that connects field equipment such as sensors, actuators and controllers to a central controller. The bus functions as a Local Area Network (LAN) for the field equipment. The specification recommends a maximum length of 1.9 km (6,234 feet) for Type A cable. Up to four repeaters can be used to extend this to five times this length.

The data highway is a communication bus that permits the distribution of the control functions. Data highways vary in length and can be in excess of 5 miles. Most longer highways use coaxial cable including Twinax cable which is a twisted and shielded coax. Shorter highways are often constructed from twisted and shielded copper pairs. Fiberglass cable or fiber is popular for point-to-point connections.

Fiberoptics eliminates problems from electromagnetic and radio frequency interference, ground loops, and common mode voltages. It can carry more information than can copper conductors and is lighter and easier to handle than coaxial cable.

The devices producing and receiving signals must be compatible with the highway transmitting the information. This information is in the form of binary-coded numbers called bytes.

A device is compatible with a data highway at the physical level if it can be connected to the highway and can produce the correct sequence of binary ones and zeros for both the transmit and receive modes.

The information must be encoded according to the bus or data highway protocol. This is the set of rules for moving data across the communications link. One protocol that has been widely used for distributed control system highway is High-level Data Link Control (HDLC). Other protocols used in distributed control systems include the IBM Binary Synchronous Communications (BISYNC), Digital Data Communications Message Protocol (DDCMP), and Synchronous Data Link Control (SDLC).

Device Access

Besides providing physical and message format compatibility, it is necessary to establish which device can transmit. This access must be long enough for loop dynamics and alarms to function properly.

There are several assess used in distributed control highways. One technique uses a traffic director to grant privileges. This is based on polling the stations on the highway or following a priority sequence.

Another technique uses a token that is passed from station to station. When a station has the token, it can communicate with the other stations. When it completes its transmitting tasks, the token is passed to the next station in the sequence.

Carrier Sense Multiple Access is used in some distributed control systems. All stations listen to the bus and any station that needs to transmit can do so, providing no other station is already transmitting. If two stations start simultaneously, control is given to the higher-priority station.

Another method is to use a shared memory in the data highway controller. This data is then available to every station. If a station needs information, the data highway controller listens for the data and places what it needs in its shared memory section. The station then uses the data from the shared memory when it needs to execute a control function.

Dedicated Card Control

The I/O for control loops, which should continue functioning even if the data highway fails, should be connected to dedicated card controllers. This is the most expensive method of PID implementation in a DCS system, but it is also the most reliable. There is a variety of card controllers having from one to twenty analog inputs and two to eight digital inputs and outputs. Execution speeds usually range from four to ten executions per second, and their structure is similar to that of a digital electronic controller.

The card controllers are usually connected to the data highway of the DCS through a data concentrator. If the central processor or the data highway fails, only the communication link is lost. The control loop will continue to function.

Another technique is the unit operations controller (UOC). A UOC is dedicated to a unit operation. All I/O associated with that unit operation is connected to that UOC. If this option is used, control is not lost if the UOC is functioning, even if the central processor or the data highway fails.

UOCs often have low-level operator's interfaces with local displays and operator access to setpoints and other adjustable parameters. A variety of workstations are available for NEMA 4 or NEMA 12 requirements with physical protection against dust, heat and humidity.

Local Multiplexers

A local multiplexer does not execute any control functions. It only transmits I/O to and from the central processor. If PID or other control functions are to be performed, they must be done in the central processor.

Smart Transmitters

These field instruments operate as standard 4 to 20 mA devices with a superimposed digital signal which is used to transport the measurement and configuration data. The transmitters are configured and interrogated on site by plugging a hand-held terminal into the supply line. The concept of a smart transmitter in this form has only a few of the advantages of a digital fieldbus.

Several protocols have been used including DE (closed), INTENSOR (open), and HART (open). HART has seen some acceptance in the United States.

HART allows the connection of several smart transmitters in a star structure. Problems with this topology include the long response times of the system which can be up to 15-seconds. There has also been the lack of a standardized method for programming and interrogating the instruments of various vendors.

Building Automation Technology

A building automation system may be a digital control system. These are packaged as Energy Management and Control Systems (EMCS) or Monitoring and Control Systems (MCS). All of these are centralized control systems with local control loops for HVAC. A basic 10-channel system can be used for one building. Systems with several thousand points are available for serving a large complex of buildings.

The local loop controls serve the building HVAC systems and the building automation system (BAS) serves the entire building or building complex. The signals which the BAS receives or transmits may be analog or digital. The following designations are used:

AI analog input
AO analog output
DI digital input
DO digital output.

The BAS can override local loop control to provide duty cycling for motor loads, demand limiting, load shedding/restoring, day-night setback, economizer or free cooling cycles and enthalpy control with cooling from outside air. Boiler and chiller sequencing and optimization can also be on override control.

Building automation also includes energy management through the following packaged systems:

Energy management systems (EMS),
Energy management and control systems (EMCS) and
Facilities management systems (FMS).

The simpler EMS usually provides time-of-day programming, duty-cycling, demand load control and space temperature monitoring for a building or group of buildings with about 40 control points. The more complex EMCS allows centralized control of the HVAC systems in a building or a building complex with software-based control for duty cycling, demand limiting, day-night setback, economizer and enthalpy changeover, ventilation and recirculation control and boiler optimization. Other functions include chiller optimization, chilled water temperature, condenser water temperature reset, chiller demand limiting, lighting control, boiler monitoring and supervision, maintenance management, and trend logging. Fire and smoke detection and building security functions may be included as well. Management functions, such as budgeting, cost accounting, and building maintenance program are usually included in a facilities management system (FMS).

Local Loop Control

The BAS will usually impose setpoint control on the local loop control system. The control point adjustment is implemented using the input from an analog sensor. Besides changes in operating setpoints, control position adjustment is implemented by using sensors from the controlled device to operate the control loop.

Scheduled Start-Stop and Cycle Operation

A scheduled start-stop software program will start and stop the HVAC system equipment on a programmed schedule based on the time of day and day of week. This provides energy conservation by turning off equipment or systems during unoccupied hours. A feedback signal is used to indicate the on-off status of the controlled equipment. This feedback signal verifies that the start or stop command has been carried out. An alarm will indicate an equipment failure or a local start or stop operation.

Optimized Start-stop

Optimizing the starting and stopping is done by automatically adjusting the equipment operating schedule based on measurements of the space temperature, outside air temperature and humidity. In a scheduled start-stop program, the HVAC systems are restarted prior to occupancy. This is done to allow enough time for cooling or heating to take place and does not consider the space and outside temperatures.

The optimizing start-stop program uses a sliding schedule. The precool or prewarm times are calculated using a predictor-corrector technique. This allows for the thermal inertia of the structure, the capacity of the HVAC system and the current space and outside air conditions. The result is the minimum time for HVAC system operation prior to the start of the occupied cycle and the earliest time for stopping the HVAC equipment at the end of the day without losing control of the internal environmental conditions.

Duty Cycles

Duty cycling involves the shutting down of equipment for predetermined periods of time during normal operating hours. This function is generally applied only to HVAC systems. Duty cycling is possible because most of the time HVAC systems do not operate at full load. The system can be shut down for a short period of time and there will not be an excessive temperature drift or comfort change during the shutdown period.

Generally, duty cycling will reduce the volume of outside air using an outside air intake damper under local loop control. The airhandling systems is cycled off for a period of time out in each hour. This is typically 15-minutes for each hour of operation. The off time period is under program control and is decreased when the space temperature conditions are not satisfied, and increased when the space temperature conditions are satisfied.

If the space temperature and humidity conditions cannot be maintained during the off period, the program may use a fairness scheme where the temperatures in all areas will deviate equally from comfort levels.

Duty cycling can be used in conjunction with demand limiting, scheduled start-stop and optimized start-stop programs, but it is not recommended for variable capacity loads, such as variable volume fans, chillers, or variable capacity pumps. For equipment without a variable output cooling section, a duty cycle may be used for the cooling section while the fan is kept operating.

Demand Limiting

Demand limiting involves the shedding of electrical loads to keep the electrical load below a specified peak value. This is done to prevent a new maximum electrical demand from occurring. Some utility rates are based on a maximum rate demand charge and exceeding the target demand value can cause an increase in electrical rates for a year. The metering of peak demand values may be based by the utility company on fixed demand intervals, sliding window intervals and time of day.

Complex demand limiting and reducing software programs are available. These programs continuously monitor power demand and calculate the rate of change of the demand in order to predict the future peak demand. This is done using predictor-corrector techniques. If the peak demand is predicted to exceed the target value, specified electrical loads are shut off on a scheduled priority basis.

HVAC systems are usually the first to be shut-down. If space conditions cannot be maintained, a software routine will cause the temperatures in all areas to deviate equally from established comfort levels using the available HVAC capacity. Demand limiting must operate in conjunction with duty cycling so that any one load is not cycled on or off during the wrong time or for an excessive number of times.

Day-Night Setback Control

The energy for cooling or heating during unoccupied hours can be reduced by raising cooling cycle temperature setpoints or lowering heating cycle temperature setpoints. Additional energy savings can be achieved if the program closes outside air dampers and shuts off exhaust fans during unoccupied hours. Day-night setback needs to operate in conjunction with scheduled start-stop and optimized start-stop programs.

Economizer Cooling

An economizer cycle in air conditioning systems can be cost-effective, depending on the climate and the type of system involved. An air-side economizer uses outside air to reduce the cooling requirements when the outside air dry-bulb temperature is below the mixed air temperature. A water-side economizer uses an evaporative liquid cooling system, such as a cooling tower, to cool chilled water for circulation. The economizer is effective when the outside wet-bulb temperature is about 5 to 8°F lower than the required chilled water air temperature.

The temperature at which an air-side economizer functions will depend on the outside dry-bulb temperature. When the bulb temperature is above the changeover temperature of the economizer, the outside, return and relief air dampers are controlled to provide the minimum required outside air. When the bulb temperature is below the changeover temperature, the dampers are positioned to maintain the required mixed air temperature. This is similar to an enthalpy cycle changeover, where the changeover depends on the total heat content of the outside air compared to that of the return air.

Water-side economizer changeover temperature depends on the outside wet-bulb temperature. When the wet-bulb temperature is above the changeover temperature, the water-cooled chiller provides chilled water. When the wet-bulb temperature is below the changeover temperature, the cooling tower water is switched from the condenser to the chilled water piping system or to the low temperature side of a heat exchanger in the chilled water circuit. This type of economizer cycle control cannot be used if humidity control is required, or if an air-side economizer is controlled by enthalpy changeover.

Enthalpy Changeover

Enthalpy changeover in an air-side economizer can be cost-effective depending on the climate and the type of system used. Outside air can be used to reduce cooling requirements when the total heat content, or enthalpy, of the outside air is lower than the enthalpy of the return air. If this condition is met, the outside and return air dampers are modulated by local loop controls to admit enough outside air to minimize the cooling requirements. Any unused return air is exhausted to outdoors.

If the outside dry-bulb temperature is greater than the required supply air dry bulb, the mechanical cooling system will need to operate. If the outside air enthalpy is greater than the return air enthalpy, the dampers are positioned for minimum outside air volume and the system operates on mechanical cooling. Enthalpy changeover is not used when dry bulb changeover is used for air-side economizer control.

Ventilation Control

Ventilation and recirculation control is needed when the use of outside air might cause an additional thermal load during warm-up or cool-down cycles prior to occupancy of the building. During unoccupied periods, the dampers are closed. During occupied periods, the dampers are under local loop control.

In the cooling season, if the outside temperature is cooler than the required space temperature, the outside and exhaust air dampers are opened, and the fans are turned on. In the heating season, if the outside temperature is warmer than the required space temperature, the outside, return and exhaust air dampers are modulated under local loop controls. This will admit enough outside air to minimize the heating requirements.

When the outside temperature is above the required supply air temperature, the cooling system will need to operate. In systems with enthalpy sensing, if the outside air enthalpy is greater than the return air enthalpy, the outside, return and exhaust/relief air dampers are positioned for minimum required outside air. The ventilation and recirculation routine must operate in conjunction with any scheduled and optimized start-stop programs prior to building occupancy.

Hot-Deck and Cold-Deck Temperature Control

A hot-deck and cold-deck temperature reset program is used in parallel path heating and cooling systems. These dual duct and multizone systems use a parallel arrangement of heating and cooling surfaces, which are called hot- and cold-decks. The air in the cold-deck is usually cooled by chilled water coils with an unregulated water flow or by mixed air economizer temperature control for the lowest supply air temperature. The air temperature from the hot-deck is usually controlled from an averaging bulb sensor with outside temperature control reset.

Cold and hot air streams from the decks are combined to meet individual space temperature requirements. Without any control system optimization, the system mixes the two air stream to produce the desired temperature. A greater difference between the temperatures of the hot- and cold-decks results in less efficient system operation.

The hot-deck and cold-deck temperature reset software compares the input data and determines which areas have the greater heating and cooling requirements. It then computes the optimum values for the hot-deck and cold-deck air temperatures.

These optimized cold-deck and hot-deck temperatures will give the lowest energy inputs when the cold-deck and hot-deck air streams are mixed. The inputs from space temperature sensors and mixing box or damper position sensors are used to determine the minimum and maximum deck temperatures. This software program operates in conjunction with scheduled and optimized start-stop routines.

Reheat Coil

The reheat coil reset software is used in terminal reheat systems which operate with a constant cold-deck discharge temperature. The air is heated by reheat coils in response to individual space temperature controls.

The reheat coil software selects the reheat coil with the lowest discharge temperature requirement. This is determined as the coil with the valve nearest to closed position, which indicates the lowest amount of reheat required. The program resets the cold-deck temperature upward until it reaches the supply air temperature required by the zone with the least demand for reheat.

If humidity control is required, the program will prevent the cold-deck temperature being reset upward past the point where the maximum allowable humidity is reached. In air conditioning systems with no reheat coils, the program will reset the cooling air temperature upward to the point where the space with the greatest cooling requirement is just satisfied.

Boiler Optimization

This program is used in heating plants with multiple boilers. The optimization of boiler plants depends on the selection of the most efficient boiler or combinations of boilers to meet the heating load. Boiler operating efficiency can be obtained by monitoring the fuel input as a function of heat output. This should take into account the heat content of the condensate return and make-up water. The burner operating efficiency is monitored by measuring the O_2 or CO in the boiler flue.

Chiller Optimization

The chiller plant optimization program is used in chilled water plants with multiple chillers. The program selects the chiller or chillers needed to satisfy the load with minimum energy consumption.

If the chiller is started from a warm start, the chiller capacity is held back to keep the system from going to full load. This time period allows the system to become stabilized so the true cooling load can be determined.

The analysis of trend logs can show when the heat transfer surfaces in evaporators and condensers are fouled and require cleaning to maintain a high efficiency.

Chilled Water Temperature

The energy required to produce chilled water in a refrigeration machine depends mainly on the temperature of chilled water leaving the evaporator. The refrigerant suction temperature is dependent on the chilled water temperature. A high chilled water temperature means a high suction temperature and a low energy input per ton of refrigeration.

The chilled water temperatures are set upward during non-peak operating hours. They are set to the highest value which will still meet the space cooling and humidity control needs.

Condenser Water Temperature

Another user of energy in water-cooled condenser systems is the temperature of the condenser water temperature entering the chiller.

Chiller units are usually designed for a condensing temperature that will allow refrigerant flow at full load. Optimizing the energy performance of the refrigeration systems requires that the condenser water temperature is reset downward when the outside air wet-bulb temperature allows lower condenser water temperatures. The minimum water temperature is limited by the minimum condensing temperature at a specific percentage of capacity.

The software resets the condenser water temperature downward using the following techniques:

- Operating cooling tower fans,
- Opening cooling tower dampers and
- Positioning cooling tower bypass valves.

Chiller Demand Control

Water chiller control involves limiting the maximum available cooling capacity and thus the power needed by the machine. One way to do this is to reduce the maximum available cooling capacity in several fixed steps on a demand limiting basis. This reduces the electrical demand without completely shutting down the chiller.

When a chiller is under demand limiting, a single step signal is transmitted to reduce the chiller limit adjustment by a fixed amount. The demand limit adjustment is accomplished by changing output taps of a control power transformer or the control air pressure to the chiller compressor vane operator. The step signals are transmitted until the demand limiting situation is stabilized. Incorrect capacity step control can force the refrigeration machine to operate in a surge condition, which may cause major damage. The minimum cooling capacity limit is critical and must be part of the software program. Typically, surge conditions occur in centrifugal compressors at loads below about 15% of the rated capacity.

Electrical Demand Measurements

Electrical demand measurements for demand limiting can be done in several ways, depending on the electrical utility used. One system uses a pulsed signal from the utility's demand meter which can be interpreted by the building automation system. Other techniques include current transformers and potential transformers on the incoming electric service to the building.

Diagnostics

The building automation software should include an I/O failure summary which explain the failure mode for each point. The failure modes include the following:

> FAIL ON
> FAIL OFF
> FAIL TO LAST COMMAND.

Industrial Intranets

Automation systems have been evolving in a hierarchical manner with different techniques used for the following networks:

- Enterprise – top level; uses conventional computers, with high versatility. Applications: E-mail, groupware, databases.
- Fieldbus – middle level; networks with both versatility and performance.
- Device – bottom level; high performance cyclic buses for control system remote I/O.

The fieldbus level is highly fragmented with Profibus, FIP, DeviceNet, ControlNet and MODBUS Plus and little interoperability. The interfaces between layers require data collection and gateway techniques.

A single set of networking techniques such as Ethernet and TCP/IP can be used to communicate at all three levels. The interfaces between layers can be components such as bridges and the information from devices can be visible at all levels of the network hierarchy.

Automation Networks

In the different levels of an automation network, the distinct communication networks are specialized, and the subsystems are optimized. At the device bus level, the network must collect a large number of relatively small data items with a small number of bits or bytes. It usually operates with a repetitive cycle. The speed of the cycle determines the responsiveness of the control device that needs the information. This group includes proprietary, remote I/O buses as well as simple serial multiplexing buses such as Seriplex and Interbus-S.

Seriplex

The seriplex control bus is a control network for simple control and sensing devices including pushbuttons, contactors, valves, and limit switches. A single cable connects all devices and the system can reduce installed costs by 50% to 70%. The Seriplex bus cable has a total cable length of up to 5000 feet.

Seriplex was introduced in 1990 and about 750,000 seriplex I/O points have been installed in 2000 installations. Applications include industrial control, building and facilities automation, data acquisition and even amusement parks. Systems are operating in North America, Europe, Asia, and Australia. A second generation of Seriplex technology was introduced in 1996.

The seriplex control bus transmits serially multiplexed data through a shielded 4-conductor cable using a unique physical layer. I/O data is continually transmitted in frames of up to 510 data bits, separated by short sync periods. The location of data bits in the frame corresponds to the addresses assigned to I/O devices.

The control bus operation is transparent to the sensor or control device. I/O devices can communicate directly with each other in a peer-to-peer bus mode, or with a host controller (PC or PLC) in a master/slave mode. Simple control logic can be programmed directly into the I/O devices.

The seriplex ASIC2B chip can be embedded directly in small control and sensing devices such as pushbuttons and relays. Other control networks can require interposing I/O modules. The Seriplex protocol can be used for data from 1 to 64 bits in length. Many control networks are only efficient for data lengths of 16 bits or greater which makes them ill-suited for simple control devices.

The seriplex control bus is highly deterministic, so the response time is always known and consistent. Other control networks that cannot guarantee the response time, require the use of time-consuming polling techniques to force the response time to be predictable.

The seriplex control bus allows a fast response to control events with update times of less than 1 ms for 128 I/O points. Message-oriented or overhead-intensive protocols can respond much more slowly even with baud rates of 5 to 10 times higher than that of the seriplex bus.

The Seriplex control bus can support more than 7000 I/O points which are connected through more than 300 physical network nodes. Other control networks are limited to about 4000 I/O points distributed among 32 or 64 network nodes.

Addressing

Using address multiplexing, the seriplex data capacity can be expanded to the equivalent of 960 bytes of data. The non-multiplexed capacity is 255 input bits and 255 output bits. Each bit corresponds to at least one discrete device. Multiplexing accommodates up to 3840 inputs bits and 3840 output bits, which allows up to 240 16-bit analog inputs and 240 16-bit analog outputs. By comparison, DeviceNet can accommodate up to 512 total bytes of data with a system.

The 7680 I/O bits can be updated in about 83ms. Since the multiplexing is performed on individual words, the response time of critical signals can be even faster. Updating 63 inputs and 63 outputs can be done in as little as 0.72ms in peer-to-peer mode and 1.36 ms in master/slave mode.

A seriplex system can operate at up to 98% efficiency, so up to 98% of the time is used for transmission of the actual data bits. The effective throughput is then 98kbits/s at a 100KHz data rate.

Analog Signals

Seriplex can accommodate up to 480 16-bit analog signals. The address multiplexing accommodates up to 3840 input bits and 3840 output bits, which allows up to 240 16-bit analog inputs and 240 16-bit analog outputs. Analog signals can be intermixed with discrete signals, but the analog signals must usually line up on 16-bit word boundaries. The 480 analog signals can be updated in about 83 ms.

A single seriplex device may transmit data up to 240 bits long. There is no limit to the amount of bits which can be assigned to a single device, up to the maximum size of 240 bits in a single frame of data. DeviceNet allows messages from 8 to 64 bits long, with provisions to concatenate several messages.

Noise Immunity

Noise immunity is aided by the following:
- A relatively low operating frequency.
- The noise must coincide with a clock pulse in order to disrupt data.
- A 12V amplitude pulse with 4.5V hysteresis is used to swamp out the noise level.

The second generation seriplex ASIC2B chip includes digital debounce and data echo features for noise filtering and data verification.

The network cable provides network power, which is usually isolated from I/O power. Since the noise is not directly coupled into the network cable, I/O faults will not disrupt bus operation. The network cable is shielded to the prevent electromagnetic coupling of noise signals.

Overhead

The serial-multiplexing protocol allows identification of data by its location within the data frame, rather than requiring transmission of source and destination addresses. By concentrating on I/O data transmission instead of general-purpose messaging, the seriplex protocol does not need to transmit and decode opcodes.

It updates the I/O data regularly, so collision resolution, message-prioritization and requests for data are not needed. Signal-receiving devices such as valves and contactors can use digital debouncing where the data must be stable for multiple readings before it is acted upon. Signal-transmission devices such as a host controller can use the data echo feature to verify that a remote device has correctly received its data. Multi-bit data such as analog signals can be validated using a complementary data retransmission (CDR) feature supported by Seriplex version 2 CPU interface cards and I/O devices.

If single bits of information are used, these are more likely to be transmitted correctly than multi-bit checksums and CRCs. CRCs and checksums are probabilistic mechanisms and cannot verify data as well as data echoing.

The data echo feature can be used to acknowledge messages and determine which devices are connected to the network. A logic XOR (exclusive OR) comparison of the original and echoed data can be used to detect data transmission faults. Seriplex host interfaces can check and report common bus faults, including loss of the clock signal and shorts in the data line.

Since the seriplex bus transmits single function data, one bit or one 16-bit value, no translation or interpretation steps are necessary. This simplicity makes seriplex data easier for device vendors implement. Support for users and vendors is provided by the independent Seriplex Technology Organization.

Configuration

The only configuration is setting addresses. Most systems operate in master/slave mode, and in this mode the on-board logic is not used. Second generation devices are configured through the network port, eliminating the need for either a separate programming port or DIP switches on small sensors and devices. If a Seriplex device fails and the ASIC2B chip is still intact, the chip's data can be read into the seriplex setup tool and written back out to another device in about 15-seconds.

Seriplex devices revert to known default states under all fault conditions. The Seriplex I/O devices continually monitor for bus undervoltage, bus fault, and clock loss conditions.

Interbus-S

Interbus-S is an open system, ring-based, distributed device network. The protocol is optimized for sensor/actuator networks. The I/O data is transmitted in frames that provide updates to all devices on the network. CRC error checking is used with diagnostics for the cause and location of errors. An embedded messaging protocol is provided for sending parameter and message data across the network.

The network uses two protocol chips. A Protocol Master chip operates the network from the controller board. (Figure 8-6) A Serial Microprocessor Interface chip links the I/O devices to the network. (Figure 8-7)

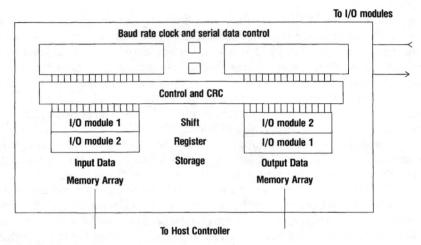

Figure 8-6 Interbus-S Controller Board

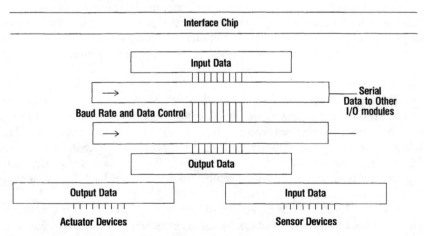

Figure 8-7 Interbus-S I/O Module

A total frame architecture is used where all data in the network is transferred in every frame. A CRC word is used at the end of the frame.

The controller board eliminates address settings by performing an identification (ID) cycle to initialize the network. This ID cycle tells the controller the type and position of I/O modules on the bus. The ID cycle is initiated by the host controller.

After the ID cycle is completed, the Protocol Master chips know the number of modules and module type present in the network. This information is available to the host controller for configuration testing.

System Data Flow

The I/O modules are connected together in a ring. The process data circulates in a continuous loop through the modules, until the shifting brings the data to the module it is intended for. After passing a CRC test, the data becomes valid in both the I/O modules and in the controller board.

During a scan cycle, the data words are shifted through the network. A data transfer requires 16 clocks. All input and output data is updated at the same time.

Scan Cycle

The scan cycle times are deterministic and depend on the number of data words. A telegram message puts the network into the scan cycle mode. The input data words from the sensors are placed into the serial bus bit stream. The Protocol Master chip sets up the output words bound for the actuators and starts shifting the data. This shifting continues until all data has moved through the network. Next, the latch phase begins where a CRC error test is made. Then, the data is available for use by sensor/actuator devices.

Network Implementation

The Interbus-S serial protocol can be implemented as a remote and local bus. Each bus type carries the same signals, but at different electrical levels. The remote bus is used for the transfer of data of up to 1300 feet. No power passes through the remote bus cable, which operates at RS-485, 500 K baud, full duplex.

The local bus connects the I/O modules to the remote bus. A bus terminal module translates the remote bus signals to local bus signals. The local bus cables are short and intended to be used within an enclosure. TTL voltage levels are used and the bus operates at 500 k baud, full duplex.

PCP Protocol

The Peripheral Communications Protocol (PCP) is used to send initialization parameters to intelligent I/O devices. The PCP protocol is a connection based, object oriented, client/server type of protocol.

When a connection is established, the client and server exchange information about the types of objects available. Object types include bytes, words, ASCII characters and arrays.

Device Networks and Enterprise Networks

The information on a device bus is usually made up of a small number of messages, with one per drop for a multidrop remote I/O network or even a single message in networks like Interbus-S.

In device networks, one of the most effective techniques is to transfer the information in a continuous scan cycle rather than transmit on change. The cycle times can be 1 ms to 100 ms.

The enterprise network is primarily used to transfer large units of information on an irregular basis. Examples include electronic mail messages, downloading Web pages, database queries, printing and transferring programs from file servers. The throughput or the bandwidth of the network is more important than the time required to transfer a data item. The network must be able to support wide variations in load and degrade by slowing up all traffic equally if overloaded rather than lockup. Since most transactions are human-initiated, the response can be within 100-200 ms.

Between these networks is the fieldbus level. The attempt here is to mix routine scanning of data with an on-demand signaling of alarm conditions, along with the transfer of large data items such as control programs and reports.

The networks used at this level, include Profibus, FIP, MODBUS Plus, DeviceNet, Fieldbus Foundation H-1. The characteristics of each of these is different enough to make seamless interconnections difficult. All of these networks have their own techniques for addressing, error checking, statistics gathering, and configuration. This complicates things even when the data itself is consistent through the networks.

Domains

One technique that can be used to improve this situation involves dividing the information at each layer into domains. The devices which interconnect these domains are made responsible for translating requests for information.

A PLC might use its device bus to scan input values and then make a subset of them available as data points on the fieldbus. Data points from fieldbus segments can be selected and made available on the Enterprise network. The devices which interconnect the domains act as gateways or data collectors.

The evolution of Ethernet networks is going from a shared to a switched design. Network switches called bridges selectively forward packets from one network to another without changing their content. This is based on whether the destination is known to be found on the other network.

These network switches allow approximately the same response as a fieldbus or device bus. This is done by controlling the traffic demand with switches and the use of client-server transactions and traffic filtering.

The devices which interconnect the levels do not need to be configured as application gateways. They are simple bridge/switch devices which have minimal system knowledge and can be easily replaced in the event of problems.

The use of the Internet and World Wide Web is allowed by the connection to a TCP/IP network. Besides Ethernet, other TCP/IP compatible networks can be used. High bandwidth backbone connections can use FDDI, long distance links can use X.25 and office connections can be made using Token Ring.

TCP/IP has a relatively large message overhead compared to the control networks so the data size which can be efficiently carried is also larger. A lower level network such as MODBUS will limit the amount of data transported in one transaction to about 200 bytes in each direction with an overhead of about 150 bytes.

If there are 10 stations and the total amount of plant data to be exchanged each cycle is 1000 bytes, the network traffic is

10 X 150 + 1000 = 2500 bytes per cycle

Multiply this by the number of scans per second to get the demand load on the bus. A scan every 100 msec would be 10 scans per second for a demand load of

10 x 2800 = 28000

Based on Table 8-9, a 10 Mbps shared Ethernet will handle this load. If the number of stations is increased or the scan time reduced, a higher speed Ethernet network may be needed. The load is the sum of all network activity which must pass through that section.

Table 8-9 Ethernet Speeds

Capacity	Type	Rate
Shared	10 Mbps	120,000 bps (about 10% bandwidth)
Switched	10 Mbps	600,000 bps (about 50% bandwidth)
Shared	100 Mbps	1,200,000 bps (about 10% bandwidth)
Switched	100 Mbps	6,000,000 bps (about 50% bandwidth)

Non-repetitive Traffic

This simple calculation works for repetitive scanning traffic. If non-repetitive traffic is allowed on the same network segments without restriction, the chance of a data item failing to get through within one scan will increase.

A limit can be placed on the devices capable of originating event-driven messages. The rate of generation of these messages is limited to a local maximum rate.

A PLC program used for alarm messages might be limited to one transaction per network scan. Then the traffic generated can be considered as if it were an additional one transaction per scan in the calculations.

Another way to control peak loading is to use a device which forwards the messages and can impose delays. This type of device is called a throttling router. MODBUS traffic is sent in the form of a transaction consisting of a request and a response. Any delay in either the request or response has the effect of delaying the transaction.

These delays will have a direct flow control effect. This technique makes more effective use of the network since it is more flexible. By manipulating the delays when forwarding the requests and responses, the non-repetitive traffic can retain the maximum throughput without compromising the network for repetitive traffic.

TCP was designed to respond to congestion situations found in a wide area network. The use of a throttling device allows even bulk file transfers between computers, which normally have a crippling effect on contention-based networks.

Web Browsers

If the PCs and control devices are connected in a seamless Internet-compatible network, then Web servers can be used to make information available in a more readable form than a MODBUS type data representation.

One technique is to let the control device act as its own local Web server. The information accessible to the device can be reported as a set of Web pages and can be accessed on any PC on the network using a Web browser. A PC computer can also act as a Web server and gather data for Web page requests.

The requests are sent out over the network and can interrogate the control devices directly if they have TCP/IP interfaces. A protocol converter or gateway would be needed without the TCP/IP interfaces.

When the Web pages are hosted on a machine which is not the true control device, the machine must relate the data to the control device and a configuration database may be needed. The database is updated as the application programs change.

The Web server can be self-configuring on installation. This can be done by using a network computer as the server. A network computer will install itself from a server on the network when it is powered up.

Figure 8-8 shows how this could work. Switches are used to prevent general-purpose file transfers and other traffic from interfering with the performance of the subnets. Communication between the control devices and the subnet are subject to throttling control. Web pages for operator control and monitoring are served with a web server on the subnet. The web server gets its configuration, directory, and pages at startup from the master copy on an Enterprise network server.

The directory of data items and reference addresses can follow the LDAP (Lightweight Directory Access Protocol) standard. LDAP can maintain a set of plant points with tag names, data types, input limits and units.

Many programming tools have a tag name export facility which can be used to generate the LDAP data. Otherwise, the tags must be manually entered.

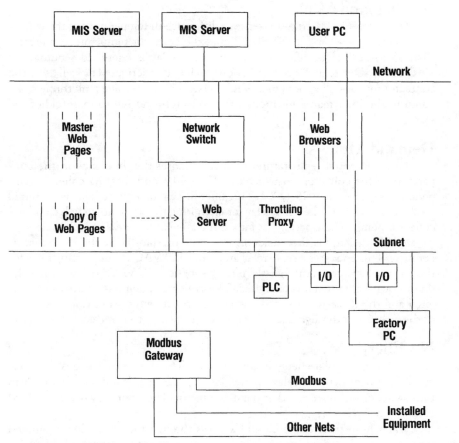

Figure 8-8 Web Server Operation

No Demand Mode

The differential thermostat senses a difference in temperature of the collectors versus storage tank at 20 +/- 3°F and the collector temperature is greater than 35 +/- 2°F. The differential temperature relay is energized sending 120 VAC to the slave relay for the drain down valve. It is activated as well as by the contactor for the collector pump. Water flows from the storage tank through the pump to the inlet manifold through the collectors and return manifold of the storage tank.

Demand Mode

The differential temperature thermostat senses the difference in the temperature of the collector versus tank at 20 +/-30°F and that the collector temperature is greater than 35 +/-2°F. The differential temperature relay is energized sending 120 VAC to the slave relay for energizing the drain down valve and the collector pump. These actions so far are the same as for the demand mode.

A heat demand is then made in the room thermostat by the closing of its contacts. This energizes a slave relay and sends 120 VAC to the starting relay for the airhandler pump. Another slave relay supplies 120 VAC to assure that the drain down valve remains energized. The collector pump contactor is deenergized and the three-way valve is energized. This reroutes water from the outlet of the airhandler through the collectors and returns it to storage.

No Energy Mode

In this mode solar energy is not available. The three-way valve is not energized. The heat demand relay is energized by the thermostat switch which in turn switches on the airhandler pump contactor. The drain down valve is held closed by a slave relay.

Water from the storage tank now flows through the airhandler pump and through the airhandler heating coils. It is returned to the storage tank by way of the de-energized three-way valve.

Appendix A: List of Abbreviations

A
A. . . . . ampere
AC . . . . ac, or a-c alternating current
A/D. . . . . Analog-to-Digital
ASCII . . . . American Standard Code for Information Interchange and is used to transmit data between computers and peripheral equipment. Each character requires one byte of information.
ASHRAE . . . . . American Society of Heating, Refrigeration and Air Conditioning Engineers
AWG. . . . . American Wire Gauge

B
bps. . . . . bits per second
BTU . . . . British Thermal Unit **(= 1054 J)**

C
C. . . . . Celsius degrees of temperature
C. . . . . centi, prefix meaning 0.01
CAD/CAM . . . . Computer-Aided Design/Computer-Aided Manufacturing
cal . . . . calorie **(gram, = 4.184 J)**; also g-cal
cfh. . . . . cubic feet per hour
cfm . . . . cubic feet per minute
CNC . . . . Computer Numerical Control
CO . . . . Carbon Oxide
CPU . . . . Central Processing Unit of computer. The microprocessor that provides processing capability, or intelligence, for a microcomputer.
CU . . . . Copper

D
D/A . . . . Digital-to-Analog
dB . . . . decibels
DC or dc . . . . . . direct current
DDC . . . . Direct Digital Control
d/p. . . . . differential pressure
DPDT . . . . double pole double throw (switch)

E
EMF. . . . . Electric Magnetic Fields are invisible lines of force that surround any electrical device.

F
F . . . . Fahrenheit degrees $[t_C = (t_F - 32)/1.8]$

G
G . . . . giga, prefix meaning 10^9
GHz . . . . Giga-Hertz
Gnd . . . . Connection to earth or another connecting body which is connected to earth.
GPM or gpm . . . gallons per minutes (= 3.785 lpm) gram

H
HP or hp . . . . . . Horsepower (U.S. equivalent is 746 W)
HVAC . . . . Heating, Ventilating, and Air Conditioning
HZ . . . . Hertz, symbol for derived SI unit of frequency, one per second (l/s)

I
IC . . . . Integrated Circuit
IEC . . . . International Electronics Commission
IEEE . . . . Institute of Electrical and Electronic Engineers
in. . . . . inch (= 25.4 mm)
I/O . . . . Input/Output
ISO/OSI . . . . . . International Standards Organization, Paris, which developed the Open System Interconnection (OSI) model.

J
J . . . . joule, symbol of derived SI unit of energy, heat or work, netwon-meter, N-m

K
k . . . . kilo, prefix meaning 1000
kBaud . . . . 1000 events or bits per second
kbit/s . . . . kilo bits per second
kg . . . . kilogram symbol for basic SI unit of mass
kHz . . . . kilo cycles per second
km . . . . kilometers
kPa . . . . kilopascals
kW . . . . kilowatts
kWh . . . . kilowatt-hour $(= 3.6 \times 10^6 \text{ J})$

L
LAN . . . . Local Area Network
lm . . . . lumen, symbol for derived SI unit of luminous flux, candela-steradian, cd-sr

lps . . . . liters per second

M

m . . . . (1) meter, symbol for basic SI unit of length, (2) milli, prefix meaning 10^{-3} , (3) minute (temporal), also min

M . . . . mega, prefix meaning 10^6

mA . . . . milliamperes **(= 0.001 A)**

Mbps . . . . mega bits per second

Mb/s . . . . mega bits per second

MHz . . . . Megahertz

mm . . . . millimeters **(= 0.001 m)**

MMI . . . . man-machine interface

ms . . . . milliseconds **(= 0.001 s)**

msec . . . . milliseconds

mV . . . . millivolt

N

NC . . . . Numerical Control

NEMA . . . . National Electrical Manufacturers Association, a non-profit organization that establishes standard specifications for electrical equipment.

NPN . . . . Negative-Positive-Negative (transistor)

O

ohm . . . . unit of electrical resistance; also omega

or . . . . orange (wiring code color)

OSI . . . . Organization for International Standards. Originally a European consortium defining standards for networking. Defines standard formats for messages sent across system boundaries.

P

PB . . . . proportional bands

PC . . . . Personal Computer

pf . . . . picofarad **(= 10^{-12} F)**

pH . . . . acidity index (logarithm of hydrogen ion concentration)

PI . . . . Proportional plus integral control logic

PID . . . . Proportional, integral, derivative process control algorithm required to fine tune such quantities as temperature, pressure and flow rates.

PN . . . . Positive-Negative (semiconductor junction, diode)

PNP . . . . Positive-Negative-Positive (transistor)

ppm . . . . parts per million

psi . . . . pounds per square inch **(= 6.894 kPa)**

PSIG or psig . . . above atmospheric (gauge) pressure in pounds per square inch

Q
Q. . . . . quantity of heat in joules, J

R
RAM. . . . . Random Access Memory
RC. . . . . Resistor Capacitor
RF or rf . . . . radio frequency
RH . . . . relative humidity
ROM-DOS . . . . Read Only Memory-Disk Operating System
rpm . . . . revolutions per minute
RTD . . . . resistance temperature detector
RS-232C . . . . . . A common protocol for connecting microcomputer system components

S
SCR. . . . . silicon controlled rectifier
SPDT . . . . single pole double pole throw (switch)

T
TC . . . . Temperature Control
TCP/IP . . . . . . Transmission Control Protocol/Internet Protocol
t. . . . . (1) ton (metric, = 1000 kg); (2) time; (3) thickness

U
URL . . . . Universal Resource Locator

V
V. . . . . volts, symbol for derived SI units of voltage, electric potential difference and electromotive force, watts per ampere, W/A
VAC . . . . volts, alternating current

W
W . . . . watt, symbol for derived SI unit of power, joules per second, J/s

Appendix B: Internet Resources for HVAC Control

Use your favorite search engine to find these Internet sites for additional information on HVAC control technology.

Contractors – Consultants

Abacus Consultants – WA
AC Contractor – Malaysia
ACME Engineering Company
AC Services
Action Mechanical
ADI htg.clg. – IL
Advance Alt.Engineering Corp.
Air Condition Specialties
Air Design Company
Air Ideal Inc. – NY
Air-Rite Air Conditioning Company
Air Systems – CT
Airtron Heating-Air Cond.Co.
Allied Heating
Allmakes Htg.Air. – NY
All Seasons Htg.Clg. – IN
All Secure – home automation
All Systems Go! Company
All Systems Heating
Alltemp Products Co. – Refrig and parts, Ontario
Ambthair Services
American Comfort Htg. – IL
Anderson Electric – home automation
ANDGAR Corp. – WA
Answer Heating – Cleaning – Teletrol
Apex Industries
Arctic Air Company
Arista Air Conditioning Company
Armco Technologies – IAQ Specialist
Ashcraft Mech.
Austin Heating – Air
Benson's Heating and Air – FL
B.E.S.T. – bldg.engr.serv.tech
Beverly Heating Clg. – PA
Bleacha and Son Refrig.
BMD Technologies – Controls

Boilerman – boilers
Brennan Associates – pm
Brennen and Associates
Champigny Plbg.Htg. – MA
Chilltrol
Clark Heating – Air Conditioning – Canada
Climate Control Company – Alerton Technical Controls
Climate Engineer
Cloud Aire – comfort accessories
Comm.Air Tech Htg.Clg.
Commercial Air-Tech
Consult-One service contracting – tracking, software – W95
Control Service of Minnesota
Cool Joe's
Cornerstone Heating – Air
Creative-Control-Design
Custom – home automation
Cypress Comfort Cooling
DADE Services
DDP – Controls
Denman Consultants
D-H Climate Control Company
Digital Energy
Dunn Mech. Service
Eco Temp Engineering – boilers
Energy Management Consultants
Energy Management Group
Enviroflex – clean room design
Enviromatics – home automation
EUA-Cogenex.Energy Consultants
Evenflo – NJ
Event Controls – home automation
Exclusive Air
Fahrenkamp Air
Florida Heating – Air
Fresh Air Services – HVAC
FSI Consult.Engrs.
Fullman Company
Gard Analytics Consultants
Gartner Refrigeration
Geocities Ems
Gold Mech Services
Goldmedal Services – HVAC
Hannabery HVAC – PA

Hernandez Htg.Clg.
Houston Boilerworks – boiler cleaning
Hungerford Mechanical
Ideal Service – TX
Indoor Climate Company
Indoor Comfort Control – CA
Ingvar Nielsson Refrigeration
JB Rodgers Mech.
J-H Larson Company
Jim's Cooling-Heating Company
J Rasa Design Consultants
KEH Energy Engrs.
Kel-HVAC
Kramer HVAC
L-H Airco Company
Linford Htg.Clg – CA
Mainstream Engineer
Management Recruiters Consultants
Margo Air – Toronto
Mastervoice – home automation
McBride Services
MEPS Assoc. – Consulting.Engrs.
Merrick Mechanical – process piping-boilers-vessels
Milner Service Company
MSI – Mechanical
Murray Company
Mussett, Nicholas & Associates
Neil and Gunter Consultants
Nth-Sintef Refrig.
Oasis Htg. Clg.
Perfection Home Systems
Perfection Mechanical
PMA – Preventative Maintenance Assoc.
Pritchard and Fallin – VA
Proplan – Bldg.Control Consultants
Pure Air Control
Rieck Mechanical – OH
Riello – oil burners
R Kraft – clean room services
Ronald R. Thomas, PE
Saturn Enterprises
Scott Plumbing Heating
Secant – home automation
Sippin Company – CT

Smarthome Automation
Snyder Group
Standard Control Systems – on-site commissioning
Standard Electric – MI
Standard Refrig.
Swauger Services
Thermalrider
Thomas Consulting Engineers
Tom's Heating
Troy Boilerworks – online help
Tsuchiyama, Kaino & Gibson
Tufarella Mechanical – NY
United Systems – Teletrol
Unitize Company – Ohio – rigging, piping, fabrication
Unitize Mechanical
University Air and Heating
Vintage Air Company – AK
V-Tech Inc – home automation
Walker Cooling-Heating Co.
Waterloo-Ozonair Ltd. – offshore htg.clg.
William Henderson Plbg.Htg. – NY
Wong Brothers Refrigeration – Malaysia
X-Energy – home automation
Controls – Systems
Accu Aire – Laboratory hood/control
Adsett Associates Data Acquistion – Control
AET – Andover
Allen-Bradley
Altech Controls
Altech Controls – Refrig.Defrost
Amtrol – access-security
Andover Controls
Argus Controls – greenhouse
Automated Building Controls
Automated Logic
Automatic Environment Controls
Automation Controls Inc. – municipal
Automatrix
Avion Controls – OEM automation
BAPI – Bldg.Automation Prod.sensors
Bayley Controls
Belimo-Actuators
Boston Edison Service – controls
Bromley Instruments – flue gas controls, industry

Building Automation
California Economizer
Cambridge Accusense – airflow
Cli-max Controls – Automation
Cline Co. Electric Components
Cognitek – Automation
Computer Perf.Upgrd.-Speech Recog.
Consolitech Corp. – notebooks.touchscreens
Controller Equipamentos – Brazil
Control Logic System
Control Loop Tuning – Tech Paper
Control Systems Tech
Control Tech
Crabtree and Associates – PLC
Cristal Controls Lonworks Devices
CSI – Controls
Dak Powerhouse Co.
Datalab Inc.
Delta Control Products – valves
Delta Controls – Intelligent Building Automation Systems
Delta Level Controls
Detroit Controls
Druck Pressure Controls
Dwyer Institute
Ebtron Inst.Co. – airflow
Echelon Lonworks – DDC network interface
EDA Controls – Amon.gas det.
EIL Instruments – Refrigeration
Elmwood Sensors
Elysium – power condition
Enco Controls – andover
Energy Experts – controls
Energy Management – andover
Energy Management of New England
Energy Systems Lab.
Enertel Controls
Engineered Automation System
Entel – automation
Environmental Tech
E.R.Mertz – signal condition
Eurotherm Controls
Foxboro Co.
Fuzzy Environmental
G-Controls Rep.

General Eastern – humidity sensors
George Lynch – Controls
Graham – Variable Speed Drives
Greystone Energy System – sensors
Harris Controls – utility
HB Instruments
Heat Timer – controls
Helix Controls for Herpotologists
Holt Electric Motors – repair
Home Automation
Honeywell
Hycal Sensors
ICAMS – Automation software
IDC – Variable Speed Drives
IDM A-C Drives
Innotech – controls – AU
Instrument Technology Inc. – scopes
IPC Process Control
Johnson Controls
Kele Controls – sensors, components
KMC Sales – controls
Landis Gyr.
Landis Gyr. Danish
Landis-Staefa
LK Controls – control repairs
Magnetek – Variable Speed Drives
Metalware Technologies – BAS software solutions
Micro Pneumatic Logic – OEM
Minco Sensors
MM Control – valves, regulators, process
Motortronics – motor controls
MPL – press.sws.
Nerger Lahr – Motion Tech., Inc. – stepper motors
NW Air – Delta Controls
Omron Controls
Pacific Enterprise – controls
Paragon Technologies
Pensar Corp. & Lonwords Technology
Phoenix Controls – Laboratory hood/control
PLC Links
Richard Zeta-Building Automation
Rolbit Manufacturers – thermostats
Rotork – Actuation
Saftronics Motor Controls

Samson Controls Inc. – process
Savastat – Boiler controller
Sensor Magazine – sensors
Setra – sensors
Siebe Environ.Controls
Smart Concepts
Smarthome Automation
Solidyne Controls
Sphere Systems
Sprague Co. – the capacitor people
SSAC – Relays, Electric – OEM components
Staefa Company
Staefa Control System
Sylva Controls – PLC
Sylva Energy Systems
Symcom Controls
Symcom – current xfmrs
Symcom Inc. Motor protection
TA Controls
TCS-Baysys Controls
TC Technologies
Teletrol – controls
Temperature Control Co.
Texas Systems – controls
Thermodisc Co.
Toshiba
Toshiba – Variable Speed Drives
Triad Technologies – remote control
Triatek – controls
TSI Sensors
TwinCo Supply Corp.
Unity Systems
USA Manufacturers
US-Energy Controls – Automation
Vaisala Institute Co.
Watt Stopper Inc.
Weed Industrial Co. – Temp.sensors
Weed Institute – Sensors
Westinghouse
WK Hile – controls representative
Wonderware – Automation Software
WSC HVAC Controls
Zcnx HVAC Controller
Zener Motor Controls

Control Technologies
Advantage Instruments
Intelligent Controller Systems
PLC Open
PLC Tutor
Programmable Logic Controllers
Programmable Logic Jumpstation
SRI's Artificial Intelligence Center
Toshiba PLC Controllers
Device Networks
(General Technologies)
ACCESS bus
ACORN 1479
Actuator-Sensor Interface (AS-I)
Anybus
ARCNET
Austrian Centre of Excellence for Fieldbus Systems
Automation and Process Control
BITBUS
Controller Area Network (CAN)
      An Introduction to CAN
      Automation Process and Control (in German)
      CANalyzer
      CAN-Bus Links
      CAN Hydraulic Users Group
      Control Area Network Sources
      Dearborn Group Technology
      Intel
      International CAN Users and Manufacturer's Group
      IPE
      Kari Ratliffe's CAN links
      Ken Tindall
      KVASER
      Motorola
      Newcastle University Robotics and Automation Group
      University of Warwick CAN Centre
Consumer Electronics Bus (CEBus)
      CEBus Industry Council (CIC)
Device Net
      Open DeviceNet Vendor Association
DIN-Measurement Bus
Distributed Intelligent Actuators and Sensors (DIAS)
Ecole Polytechnique Federale de Lausanne's Fieldbus Page
Ethernet

European Installation Bus
European Intelligent Actuation and Measurement User Group
FICIM
FICOMP
Fieldbus: A Neutral Instrumentation Vendor's Perspective
Fieldbus Forum
Fieldbus Foundation or Foundation FieldBus
Fieldbus Home Page (France)
Fieldbus in Industry: What is at stake?
Fieldbus International (FINT)
Fieldbussysteme (in German)
Fieldbus Tutorial
FieldComms International
FIP/WorldFIP
Foundation Fieldbus
    Dow Fort Saskatchewan (first FF application)
    Fieldbus Foundation
HART
    Hart Communication Foundation
    Virtual HART Book
12C Bus
INFIDA
InterBus-S
    InterBus-S (in German)
    InterBus-S Club
    Phoenix Contact's InterBus-S Site
Inter-Control Center Communication Protocol (ICCP)
Local Operating Network (LON)
    Echelon's (LON) Site
    Flexnet-Action Instruments
Maritime Information Technology Standard (MITS)
Manufacturing Message Specification (MMS)
MODBUS
Multiple Master, Multiple Slave (M3S)
NRD/TCP
Open Low-Cost Time-Critical Wireless Fieldbus Architecture
(OLCHFA)
    OLCHFA research effort
    OLCHFA status in industry
Permormative Requirements for Intelligent Actuation and
Measurement (PRIAM)
P-Net
    International P-Net Organization
    PROCESS-DATA Silkeborg ApS

Profibus
>Profibus Description (in German)
>Profibus International

Programmable Fieldbus Cell (Proficell)
Semiconductor Equipment Communication Standard (SECS-I/II)
>Semiconductor Equipment and Materials International
>Sensoplex
>GW Associates Sensoplex

SerCOS
Seriplex
Smart Distributed System
>Honeywell
>Specifications

Smarthouse
Steinhoff Automations and Fieldbus Systems
Synergetic Microsystems' Fieldbus Comparison
Universal Serial Bus (USB)
>Computer Access Technology Corporation
>USB Implementer Forum

X10
>Home Automation Systems, Inc.
>Page for comp.home.automation

Education Institutions
Arizona State University – Chemical, Biomedical, and Materials Engineering (USA)
Auburn University – Chemical Engineering (USA)
Brighham Young University – Chemical Engineering (USA)
California Institute of Technology – Chemical Engineering (USA)
California State University – Chemical Engineering (USA)
Cambridge University – Engineering (UK)
Carnegie Mellon University – Chemical Engineering (USA)
Case Western Reserve University – Chemical Engineering (USA)
City College of New York – Chemical Engineering (USA)
Clemson University – Chemical Engineering (USA)
Colorado State University – Chemical Engineering (USA)
Columbia University – Chemical Engineering, Material Science and Mining Engineering (USA)
Cornell University – Chemical Engineering (USA)
Dartmouth College – Thayer School of Engineering (USA)
Drexel University – Chemical Engineering (USA)
Edinburgh University – Electrical Engineering (UK)
ESIGEC Chambe'ry (in French) (France)
Florida State University Dept. of Chemical Engineering (USA)
Georgia Institute of Technology – Chemical Engineering (USA)

Glasgow Caledonian University Centre for Industrial Bulk Solids Handling (UK)
Griffith University Intelligent Control Systems Lab. (ICSL) (Australia)
Helsinki University of Technology (Finland)
Illinois Institute of Technology Armour College of Engineers (USA)
Institute of Paper Science and Technology (USA)
Iowa State – Chemical Engineering (USA)
Johns Hopkins University – Chemical Engineering (USA)
Kansas State University – Chemical Engineering (USA)
Kansas University – Chemical and Petroleum Engineering (USA)
Kings College, London, Division of Engineering (UK)
Lehigh University – Chemical Engineering (USA)
Loughborough University – Manufacturing Engineering (UK)
Louisiana State University – Chemical Engineering (USA)
Massachusetts Institute of Technology (MIT) – Chemical Engineering (USA)
Michigan State University – Chemical Engineering (USA)
Middlesex University Advanced Manufacturing & Mechatronics Centre (UK)
New Jersey Institute of Technology – Chemical Engineering (USA)
New Mexico State University – Chemical Engineering (USA)
North Carolina State University – Chemical Engineering (USA)
Northwestern University – McCormick Chemical Engineering (USA)
Ohio State University – Chemical Engineering (USA)
Ohio University – Chemical Engineering (USA)
Oklahoma State University – Chemical Engineering (USA)
Oregon State University – Chemical Engineering (USA)
Princeton University – Chemical Engineering (USA)
Purdue University – Chemical Engineering (USA)
Queens University of Belfast – Electrical and Electronic Engineering (Ireland)
Renselaar Polytechnic Institute – Chemical Engineering (USA)
Rice University – Chemical Engineering (USA)
Riga Technical University (Latvia)
Rutgers University – Chemical and Biochemical Engineering Graduate
             Program (USA)
South Dakota School of Mines and Technology – Chemical Engineering (USA)
Stanford University – Chemical Engineering (USA)
State University of NY at Buffalo – Chemical Engineering (USA)
Syracuse University – Chemical Engineering and Materials Science (USA)
Tallin Technical University (Estonia)
Tempere University of Technology (in Finnish) (Finland)
Tampere University of Technology (in English) (Finland)
Technical University/Budapest (Hungary)
Texas A&M – Chemical Engineering (USA)
Texas Tech – Chemical Engineering (USA)
Tufts University – Engineering (USA)
Tulane University – Chemical Engineering (USA)

University College, London – Chemical and Biochemical Engineering (UK)
University of Alabama, Huntsville – Chemical and Materials Engineering (USA)
University of Arizona – Chemical and Environmental Engineering (USA)
University of Arkansas, Fayetteville – Chemical Engineering (USA)
University of Bradford – Mechanical Engineering (UK)
University of California, Davis – Chemical Engineering and Materials Science (USA)
University of California Santa Barbara – Chemical Engineering (USA)
University of Cambridge – Chemical Engineering (UK)
University of Cincinnati – Chemical Engineering (USA)
University of Connecticut – Chemical Engineering (USA)
University of Dayton – Chemical Engineering (USA)
University of Delaware – Chemical Engineering (USA)
University of Glasgow – Mechanical Engineering England (UK)
University of Illinois at Chicago – Chemical Engineering (USA)
University of Illinois at Urbana-Champaign – Chemical Engineering (USA)
University of Liverpool – Materials Science and Engineering (UK)
University of Maine – Chemical Engineering (USA)
University of Manchester – Mechanical Engineering (UK)
University of Maryland – Chemical Engineering (USA)
University of Massachusetts at Amherst – Chemical Engineering (USA)
University of Michigan at Ann Arbor – Chemical Engineering (USA)
University of Minnesota – Chemical Engineering and Materials Science (USA)
University of Mississippi – Chemical Engineering (USA)
University of Missouri Columbia – Chemical Engineering (USA)
University of Mew Mexico – Chemical and Nuclear Engineering (USA)
University of Notre Dame – Chemical Engineering (USA)
University of Sheffield – Automatic Control and Systems Engineer (UK)
University of Surrey – Chemical and Process Engineering (UK)
University of Tennessee, Knoxville – Chemical Engineering (USA)
University of Texas at Austin – Chemical Engineering (USA)
University of Utah – Chemical and Fuels Engineering (USA)
University of Wales Cardiff Intelligent Systems Lab (UK)
University of Wales Swansea – Mechanical Engineering (UK)
University of Wisconsin – Chemical Engineering (USA)
Yale University – Chemical Engineering (USA)

Embedded Systems
BizWeb
DSP and Embedded Systems Tools
EG3: Electronic Engineering Net Resource
Embedded.com

Embedded System Conferences
Embedded Systems Programming Europe magazine
Embedded Systems Programming magazine
IEEE Real-Time Research Repository
Internet Resources
Miller Freeman Directories
Nerd World Media
Ptolemy Project
Real-Time magazine
Softaids's Embedded Web

Equipment Manufacturers
Aaon Corp. – AHU's-RTU's-heat transfer coils
Adapt INC. – spot-portable clg.
Air Cool – window ac
Air Movement Group – UK
Air Rover – spot-portable cooling
Airserco Manufacturing
Air Supply – product
AJ Manufacturing – arch.prod
ARCO Aire
Armstrong – pumps, heat exchange, hydronics
Baltimore Aircoil
Bearington – boiler sales
Berner Corp. – air curtains
Bradley-Lomas-Electrolok, UK – access, fire, condition pumps
Bry Air – desicant dehumidification
Burnham – Boilers
Butler Ventamatic – attic-exchange fans
Caldwell Energy – thermal storage systems
Calmac Manufacturing – off peak ac-refrigeration
Cargocaire – desicant dehumidification
Carnes – air distribution products
Carrier Corp.
Carrier – MidAtlReg
Certainteed Air Duct Liner
Chillers On-Line EvapCo and more
CIMCO – industrial refrigeration
Cleaver Brooks – Boilers
Climatemaster WSHP and more
Cold Point – window ac
Colony Custom Curbs
Comfortmaker
Cook Fans

Copeland Corp. – compressors
Dec-Therma-Store – heat recovery – heatpumps
Ductmate – air distribution duct connectors and access
Dupont – refrigeration
Duro-Dyne – air distribution duct products
Easco Boiler – firetubes
Elf-Refrig.R406A
Enocar Energy – geothermal heatpumps
Evapco cooling towers, dry coolers
Florida Heatpump – water source HP's
Forkor's – desicant dehumidification – ice rinks
Greenheck Fans uc
Grundfos Company – hydronic pumps
Hart and Cooley – air distribution products
Hartford Compressors
Heatway – radiant floors, snow melting
Heil
Henry Valve – manufacturer
Herman-Nelson – Portable Heaters
Hitachi Company – split system.ac
Howden Ltd. – screw compressors
Hunter Intl.Air Management
Hussman – refrigeration
HVAC Equipment – Unicom
Indeeco – electric heat
Ingersoll Rand – air hydrogen pumps
IRRSA-Industrial – shell-tube
ITI Ferrotec – solidstate cooling
ITT – fluid handling
Kathabar – dehumid.heat-recover
Kewanee – Boilers
KSB Inc. – feedpumps, industry
Kysor Industrial – refrigeration, coolers, display
L and W-haz.area equip.CA
Lennox Company unitary-resed.-comm.
Liebert Company – computer room-unitary
Magic Aire – blower coil product
Mammoth Company unitary-ahus
Maritime Geothermal-Nordic geothermal heatpumps
Marley Company cooling towers
Marlo Company ahu's-heat transfer coils
McQuay International unitary-applied system-controls
Mitsubishi Company – split system.ac
Miura Company – boilers

Mobile Control – offroad – mobile ac
Mydax Company – chillers
Myson Ltd. UK – radiators-convectors
Nautica Company – dehumidification
Nebraska Boiler – watertubes
Nordyne Company – elec.heat.manuf.housing
Nuheat Company – floor radiant heat
Peak Industries – ac equipment
Penn Ventilators – fans
Phoenix HP – heat recovery
Radiant Technologies – hydronics
Redi-Controls – refrig.recov.
Reggio – decorative grilles
Regin HVAC Product – test inst.controls
Research Products – humid air clean
Rheem/Ruud Company's unitary-resed
Ruskin Dampers
SANYO Company – ductless.split sys.ac
SEIHO – air distribution products
Snappy – Duct Products
Super Radiator – Coils
Sussman – electrical boilers
TACO hydronic pumps
Tecumseh Corp. – compressors
Tellurex Company – solidstate cooling
Tempstar
Texas cooling towers
Titus – air distribution products
Tower Technical fiberglass cooling towers
Trane Company unitary – applied systems-controls
United Thermodynamics
Ventex – louvers – dampers
Vent Tech. – flexible duct products
Water Furnace – heatpumps
York International unitary – applied systems – controls

Individual Homepages
Thomas V. Beaudoins HMPG
Stan Bunchs HMPG
Argo Cesareos HMPG
Mikes Forshaws HMPG
Russ Frenchs HMPG
George Goble Purdue Univ. Engr.
Gary Hillards HMPG for Techs

Larry James HMPG
Frank Lewangs HMPG
Rick Madrigals HMPG
Art Millers HMPG
Paul Milligans HMPG
Frank Nemias HMPG
Marc O'Briens HMPG
Steve and Pats HMPG
Steven Pettyjohns HMPG
Walter Scotts HMPG
Kevin Shaws HMPG
Keith Tomilsons HMPG
Peter Westman HMPG
Randy Wilkinson's HMPG

Intrinsic Safety
Certification of Products
Crouse-Hinds Intrinsic Safety Page
Crowcon's Gas Detection Page
Factory Mutual
Hazardous Area Approval
MTL Application Notes & Papers
Multiple Instrument Approvals
Turck's Intrinsic Safety Information

Laboratories
List of Flow Measurement Labs
National Calibration and Testing Labs (NCTL)

Links to Links – References
Aecnet – Arch.Engr.Const.network
Aha Net – Engineering – Links
Airnet – Euro
Allied Refrig. Manufacturers Links
Allied Signal – Vendor Info
Appliance Net-art of motion
Arrowhead Consumer – Information
Banner Fuel – HVAC Links
Build.Com Links
Building Online – Search Engine
Builder-Online
Building Design-Construction
Building Information Warehouse
Buildings Online – products

Buildnet
Comfort-House Consumer – Information
Comfort-Net
Connet – construction net
Contracting Business
Contractors Web
Cross Manufacturer– water filters
CS Dowling – search firm
CSN – Energy Source
Dak – Xref – HVAC Encyclopedia
Electric Utility Info
Englers Links-Euro
Energy Yellow Pages
Environmental Biz
Environmental Organization Webdirectory
Environmental Support Solutions
ESS – HVAC Mall
Facilities-Net
Fire Protection
Gasinfo – Commercial
Global Resource – eng.architecture
Home ideas – project planning
HVAC Jobs
HVAC Online
HVAC Resources – Canadian
Indoor Climate
Industry Link
Industry-Net
Inter-pro-division 15
Isdesign Net
Kirkbridge Assoc. exec.search firm
Kitchen-Net – HVAC Links
Maintenance Mall
Mr-Colorado Springs-search firm
NationalJob-Technical Positions
Netcad – HVAC Tech.Resources
Newsgroup: alt.AUTOCAD
Newsgroup: alt.HVAC
Newsgroup: alt.Lonworks
Newsgroup: sci.engr.heat-vent-ac
NW Buildnet
Oil Info Bureau
Other HVAC Jump Site – cichlid
Other HVAC Jump Site – hvac-city

Plumbing Web
Prospec-Interactive Lab
Recruiting Service
Repair-Net appliances
Save Energy.Com
Sci.engr.heat-vent-AC-FAQS
Sci-Engr-Heat-Vent-AC HVAC Jumps
Search-Placement-Firm-HVAC
Service Net
Sheet Metal Workers, AOL
Sunbelt Engineering
Sweets Division 15
Team HVAC
Techmall
Technical Assistance Bureau
Thermonet
Thomas Register of Companies
Trade Service Corp.
Water Web – equipment

MES/MRP
Enterprise Database Products
MESA International
MES: Can Your Company Afford it, Or Afford Not to Have It?
MES Implementation
MES' Pinpoint Problems
MRP

Newsgroups
(Chemical Engineering)
All aspects of chemical engineering
Chemical laboratory equipment
(Computers – General)
Computer security issues
Embedded system topics
Future of computing
Home automation
Human-computer interaction (MCI)
International standards
Laptop (portable) computers
Peripheral devices
Real-time computing
Wireless networks (moderated)
(Computers – Languages)

C language
C language (moderated)
ISO 9899
C++ language
C++ language (moderated)
C++ library/standards (moderated)
Client-side JavaScript
Object-oriented/logic
Object-oriented programming
Visual Basic
Visual Basic Databases
Visual Basic Miscellaneous
(Computers – Networking)
Cabling
Ethernet/IEEE 802.3 protocols
Neural networks
TCP and IP networks
Token-ring networks
(Computers – Operating Systems)
Arguments for/against Unix
CHORUS operating system
Connectivity products
OLE, COM, and DOE programming
32-bit Windows programming
Windows (Miscellaneous NT issues)
Windows NT advocacy arguments
Windows NT driver development
Windows NT network administration
Windows NT system administration
(Computers – Robotics)
Research in robotics (moderated)
Robotics and its applications
(Electronics)
Electronic circuit design
Elementary electronics
Fixing electronic equipment
General electronics
Integrated circuits, resistors, and capacitors
Schematic drafting, printed circuit layout, and simulation
Test, lab, and industrial electronic products
(General Measurement and Control)
Engineering of control systems
Technical engineering tasks
(Industry)

Energy, science, and technology
Food science
Pulp and paper industry
Quality
Semiconductors
(Manufacturing)
Manufacturing technology
Micromachined structure devices
Safety of engineered systems
(Scientific Techniques)
Computational fluid dynamics
Crystallography
Electromagnetic theory
Fiberoptics
Field of microscopy
Fluid mechanics
Fuzzy logic
General testing techniques in science
Magnetic resonance
Mass spectrometry (moderated)
Nondestructive tests in science
Power system protection
Science of optics

Other – Products – Support
AAA-Solar
Accu-Balance Associates – air testing
Acentech – Sound Consultants
Air Blender
Air Filters Inc.
Airhandling – duct products
Air Purification
Air Quality Management
Air Quality Management – Filters
Alack Refrig. – refrig.food service
Americlean Consultants – contaminates-abatement-monitoring
Amprobe – test inst.elec.
Armaflex – refrig.line insulation
AV Refrigeration – ice machines
Boilers and Heat Exchange
Breathe-Eze Electrostatic Filter
Brinmar winter/SOUND AC covers
Bronz-Glow protective coatings
Campanella Sound Consultants

Cannon Water Technical water treatment equipment
CH – Construction Management
Chiller Refrig. Recovery – Purge
Cleanflite Air Purifiers
Coit Services – Airduct Cleaning
Coldchain – Transport Refrigeration
Cold Equipment – Refrigeration
Cooling Systems – Biological Control
Combotronics-Fixit – Electronics Repair
Commair Rotron – centr.blowers
Cool System – heat exchangers
CS Refrigeration – New and used equipment
Custom Coils Fin-Tube Coils
Dominion Refrigeration – new and used equipment
Easyheat – Electric Heaters – Freeze protection
Efficient Combustion
Electrical Parts Site
Enco Coils A-C Reheat
Environmental Techtonics – lab simulation, testing
Envirotech – refrig.sys
Envirovac – airduct cleaning
Equipo De Refrig. – industrial sys
ETL – Testing Labs
Fasco Motors – small motor manufacturing
Findlay Equip. – refrig.food service
Flanders-Filters
Goodfellow Consultant – IAQ
Greatflame – fireplace inserts
Heatcraft – Fin-Tube Coils
Heat Pipe Technology
Hepa-Bedroom Air Cleaner
Hepworth Air Filtration
H2O Doctor – water treatment
HY-Save Refrig.Technology
IEP Personal Air Filtration
E.J. Bartell – duct products
Indhex AB – air-to-air heat exchanger
In-O-Vate Technology – dryer heat recovery box
Isolate – Hepa, Air Filtration
J and R Distributors – refrigerants
K-Mont Industry – water treatment-industry
Lecellier Wine Cellar Temp.Control
Lincoln Technology – IAQ
Little Red Book

Lordan – Fin-Tube Coils
Lordan – refrig.food service
Low Keep Refrig. – wind.pwrd.refrig.
Malco Mg. HVAC Tools
Marmat – promotionals
Masonary stove builders
Mechanical Systems – airduct cleaning
Mesi-Plant Design
Nashua – duct tape
Novare Consultants – IAQ
Nutemp Refrigeration – new and used equipment
OJ and C – Distributor.TX
Precision Aire-Filters
Purafil Filtration
Refrig.Recycling-Reclamation Service
Refrig.Supply Co. – distributors – recycling, equipment DC
Refrigeration Transfer System
Refron-refrig. solutions
Retterer Manufacturing HVAC Tools
Rocky Research – Prod.Development
Safety King – airduct cleaning – MI
Schuller Mech.Insulation
Secespol – shell 'n tube heat exchangers
Sisneros Brothers – sheet metal products
Snappy – duct products
Solar Design Associates
Solar Energy
Southern Environment – Filters
Spal Tech – small axial.centr.blowers
Star Tec – Refrig.Recycling-Reclamation
TAI-NDT Eddy Current Testing
Tarm USA – woodburning stoves
TD Air Balancing
TDrill – Tube and Pipe Fabrication Machines
Tech Instrumentation
Tei HVAC Test Equipment
Testmark Labs
Thermal Connection
Thermofoil-Insulation
Thomas Young – test & balance
Triumphe – equipment leasing
True-Cut Mg. Sht.MTL
Vicarb – heat exchangers
Ware Energy – chiller, boiler rental

Water Treatment – Appld.Science
Water Treatment Product
Watts Industries – water treatment, filters
West States – distributors
Winter-gard winter AC covers

Publications – Documents
American Machinist
Applicance Tech Talk
BBI – HVAC gas technology book
BioMedical Products
Boiler Eff.Institute – educational
Broadband Systems & Design
Building HVAC Expert
Buildings Online – Mag.
CAEnet – Computer-Aided Engineering
Center for Building Studies
Chemical Equipment
Construction Magazine
Consulting – Specifying Engineer Magazine – updated weekly
Control Loop Tuning
Cutter Information Energy Directory
Cutter Information – energy-environment
Datamation
Design News
Economy and the Environment
EEM – Electronic Engineers Master
EE Times
EDN – Electronic Design News
EDN – Products
Electronic Business Today
Electronic Design
Electronic Industry Directory
Electronic News
Electronic Packaging and Production
Electronic Products
Energy User News
Engineering News Rec.
EPA – Ozone – Refrigerants
ESS – Ozone Depletion
Facilities Management HVAC Shop
Fiberoptic Product News
Food Manufacturing
FTI-DOE building simulation software

Healthar on IAQ
HVAC – Heating/Piping/Air Conditioning
HVAC Controls & Systems-onramp books
HVAC Control – Troubleshooting Handbook
HVAC Hxchanger Magazine
HVAC – Insider
HVAC-R Directory/Source Guide
HVAC/R News
HVAC Sequences
HVAC System – webdeck
HVAC Training
HVAC Training Courses
Hydraulics and Pneumatics
Hydraulics & Pneumatics magazine
IAN – Instrumentation & Automation News
I&CS – Instrumentation & Control Systems
IC Master
IEEE – Automatic Control
IEEE – Control Systems Technology
IEEE – Industry Applications
IEEE – Industrial Electronics
IEE Instrumentation and Measurement
IEEE Micro
IEEE – Robotics and Automation
IEEE Spectrum
Indoor Air Service training
Industrial Computing
Industrial Maintenance & Plant Operators
Industrial Product Bulletin
Industry Week
InTech
ISA's Standards Department
ISA Transactions
ISA Transactions Special Issues
Laboratory Equipment
Laser Focus World
Lasers and Optronics
Lawrence Berkley – Energy-Environment
LBl – sky simulator
Lubrication in refrigeration systems
Machine Design
Manufacturing Marketplace
Manufacturing Systems
Material Handling Engineering

M&C – Measurement & Control
Means Mechanical Cost Data
Metal Heat Treating Digest
Metalworking Digest
Microwaves and RF
Modern Materials Handling
Motion
Motion Control
NAHB Research – Newsletters
New Equipment Digest
Omega Technical Training Software-Elec.
Ozone Depletion – FAQ OS-Univ
Packaging Digest
PC Computing
Personal Engineering & Instrumentation News
Pharmaceutical Processing
Plant Engineering
Pollution Engineering
Powder Bulk Solids
Power Engineer, Books, means
Plumbing and HVAC Costs
Prentiss Hall – HVAC,Books
Prestons Guides – eer.seer.eff ratings
Product Design & Development
Refrigerants – Montreal Protocol
Refrigerants – R406A
Refrig.Test – EPA certif.
Reliability Magazine – maintenance – msg.brd.
ScanTech News
Scientific Comp. & Automation
Security
Security Distributing & Marketing
Semiconductor International
Sensor Magazine – sensors
Sensors
Sigal Environment – newsletters
Steam and Refrigeration Licensing – WA
Supply House Times
Technical Software News
Technical Training Books
Tempcon Inc. – HVAC training, controls
Test & Measurement World
The Use of Ozone
Tradeshow Week

UA Local 111 – HVAC Apprenticeship
University Idaho – ground source heatpumps
Used Equipment Network
Vision Systems Design
Weather Data – Univ.Ky
Welding Design and Fabrication
Wireless Design & Development

Reference Resources
(Companies and Products)
Associates for Facilities Engineers Online
BestTest Guide Directory
BizWeb
Business Directory International Instrumentation and Controls
Channel One – the computer industry directory
Electricnet
Electronics Manufacturers on the Net
Expert Marketplace
IndustryNet Buying Guide
ISA's Directory of Instrumentation
Manufacturing Marketplace
Nell Register
Process Measurement & Control and Electrical Equipment
Supersite.net's List of Process Equipment Companies
Water Online
Web Wanderer's List of Industrial Company Home Pages
(Glossaries)
Glossary of Computer-Oriented Abbreviations & Acronyms
(Virtual Libraries)
Analyticon Information Systems
AutomationNet
Computer Information Centre
Control Components Catalog
Control Engineering Virtual Library
Designinfo Engineering Catalogs
Edinburgh English Virtual Library
ESPRIT
Flow Measurement Labs
Galaxy
Gridwise Power Guide
IEEE Computer Society's Publications
Industry Link
IndustryNet
Instrument-net

NEIS Index to EPA Test Methods
Non-stop Technical Information
Online Expo for High Technology
OPC Foundation
Readout Instrumentation Signpost
Safetylink
Systems and Control Archives
TechExpo
Thomas Register
University of Glasgow Centre for Systems and Control
Virtual Library Process Control List
Webopaedia

Reps – Distributors – Wholesalers
Action Supply Company
Addison Products – Distributors
Air Conditioning Equipment – Parts
Air Condition – Specialties
Aireco-Quemar – Refrig.AC – Dist.Can
Air Handling System Ductwork
Albina HVAC Distributors – WA
Allied Refrigeration – Dist.CA
Allied Supply Dist.
Alltemp Distributor.CA
A1-Appliance Repair – Parts
AMS Controls, sheet metal products
APCO Inc. Equip.Dist.
Aspen Htg.C.g.Dist.
Automated Bldg. Controls-NY
Automotive Air – Parts
BAPI – Bldg.Automation Prod. sensors
Bayside HVAC Product
Behler HVAC Distributors
Blackman Distributor – New York
Bronz-Glow protective coatings
Brooklyn Energy Product
Buckeye Supply Distributors and books
C and J Metal, sheet metal products
CJ Enterprises – refrig.ammon.
Climate Controls – Alerton Controls
Crescent Supply – Distributor
Dickson Supply Pumps-Heating-Water Treatment
Engel Industries, sheet metal forming
Energy Related Products – great pictures

Environmental Air Product District – OH
Familian Equip.Dist.
Faumer Co. Htg. Controls
FW Webb – wholesale Distributor
Gensco HVAC Distributors – WA
Geonex Intl. – Dist.
Ginther Sales – Wholesale Dist.TX
Global Exports
Grainger Distributor
Greentech Equip.Dist.
HTS Dist. Pumps
Hunter baseboard htg.products
ICS – Innovative Control Solutions – Delta controls
Ingle Supply
Inter-City Products
J-B Sales
Johnson Supply – Distributor
Johnstone Supply.Dist.Home
Johnstone Supply.Dist.SA-CA
Keller Supply Company
Mestek Distributors
MJC – Inc. protective coatings
National Account Sales – District
N E Thing Supply
Noland Co. – VA Distributor
O'Connor – Distributor
Park Supply Distributor
Pelonis – Ceramic Heaters
Qualitec Distributor
RIM Plbg.Htg. Distributor – New York
R. L. Shields – Delta Controls
Solco Plumbing-Heating Supply
Suburban Dynaline – heaters
Swift Htg.Elem. – elec.htg.
Tackaberry Distributor
Temperature Supply Distributors
Temperature Supply – Parts
Tempcon Inc.-HVAC Training
TESCO Distributors
Tri Cut Inc., sheet metal products
Tri-S – portable sheet metal brakes
Tuco Industrial – heaters, blowers
US – Energy Controls – Refrig.reclaim units
Wabash Power Equipment Co. – Boilers-power general

Wade & Associates – Representative
Walters Controls
Watcon-Water Treatment
Western Pacific Distributors
Yeatman Distributor

Software – CAD – Drawing – Engineering
AED-CAAD, CAD training
AUTOCAD-CAD
Best – Estimation Software
Bottom-Line-Software-Service Management
CADAIR – HVAC design software
CADKEY – 3D Design Software
Carmel Software Corp.
Cimetrics – Bacnet software.drivers
Contractor Accounting Software
C-R Technologies – thermal analysis
Diehl-Graphic CAD
Doe – Energy Software
Eley Software – energy simulation
Elite Software
Estimation Software
Etree Software
HVAC Calc. – Africa
HVAC Drafting Service
HVAC Manager – Access95 software
Idea-Graphixs
Intelicad – Autocad to plasma cutter
Intergraph-CAD
KRS Enterprises HVAC/PHC Software
Linric Co. – psychrometric software
Mc2-HVAC CAD
Memate – HVAC Software Integrators
Microworks Software – Load calc. and co$t estimating programs
Parametric Duct Designer for Autocad
Psychrometric Software
Quote – Estimation Software
Quotesoft Software
Simulation Software
STS Estimation Software
Technical Troubleshooting Software
The Specifiers – selection/spec.software
TMS – Software
Universal CAD-piping

Utility Software – steam-psychro
Vertical Market Software – acctg.disptch.jobcost
Viasoft – CAD, duct
Viasoft – cad software, duct
Visio Corp. Drawing Programs
Walker Systems Software
Wright HVAC Software Tools

Standards
American Association for the Advancement of Science
American National Standards Institute
Department of Defense
Institute of Electric and Electronics Engineers (IEEE)
ISA Standards
National Standards System Network
Society of Automotive Engineers

Tools
Process Tools
Thermocouple Accuracy

Trade Associations – Government – Institutions
ABMA – American Boiler Manufacturer Association
ACCA – Air Conditioning Contractors of America
AFE – Association of Facility Engineers
AIAA – American Institute of Aerospace and Aeronautics
API – American Petroleum Institute
ARI – Air Conditioning Refrigeration Institute
ASHRAE – Home
ASIC – American Society of Home Inspection
ASHRAE – Tennessee Chapter
ASME – American Society of Mechanical Engineers
ASTM – American Society for Testing Materials
Berkley National Lab. – Remote Building Monitoring Project
BLDG.Tech.Group – MIT
BOCA – Building Officials-Code Association
BOMA – Building Owners-Managers-Association
British Refrigeration Association
Caba-Home-Building Automation Association
Cabbet Energy Efficiency
Cabo-Council of America Building Officials
California Energy Commission
Carnegie Mellon University Environmental Research
Cobra – Cogen.Boilers.Stationary Engineers.Association

Columbia Energy – Utility
Control Systems Integrators Association
Cornell University EMCS
CPSC – Consumer Product Safety Commission
CSIA – Control System Integrator Associations
Delft University of Technology
Department of Automatic Control, Sweden
ECS – Electrochemical Society
EEC Environmental Energy Conservation UK
EIA – Electronic Industries Association
EIAJ – Electronic Industries Association of Japan
EIC – Energy Ideas Clearinghouse – WA
Envirosense IAQ
EPA – U.S. Environmental Protection Agency
EPRI – Electric Power Research Institute
ESP – Associate Energy Professionals
FETA – Federation of Enviro.Trade Assoc.UK
Fieldbus Foundation
FSEC – Florida Solar Energy Center
Gas Cooling Center
Gas Research Institute
GHPC – Geothermal Heatpump Consortium
HVAC Lab – Colorado State University
HVAC System at World Trade Center
ICS – Industrial Computing Society
IEA – International Energy Association – solar
IEC – International Electrotechnical Commission
IEEE Control Systems Society
IEEE Instrumentation and Measurement Society
IIEC – Sustainable Energy Guide
International Institute of Refrigeration
Iowa State University – Industrial Assessment Center
ISA – International Society for Measurement and Control
ISES – International Solar Energy Society
ISO – International Standards Organization
JAIST – Japan Advanced Institute of Science and Technology
MCA – Chicago
MCCAA – Mechanical Contractors Association
MSPE – Mo. Society of Professional Engineers
NAHB – National Association of Homebuilders
NASA – Engineering List
National Association of Plumbling, Heating, CLG.
National Climatic Data Center
NCMS – National Center for Manufacturing Sciences

NEMA – National Electrical Manufacturers Association
NFESC Energy Library
NFPA – National Fire Protection Association
NJHPC – New Jersey Heatpump Council
NIST – National Institute of Standards
Northeastern Univ.Bldg.Design
NPCA – National Paint and Coatings Association
NREL – National Renewable Energy Lab
NSAE – National Society of Architecture Engineers
NSPE – National Society of Professional Engineers
NW Enernet – Power and Gas
Oakland University PD Engineering Software
OK State University
PC/104 Consortium
PICMG – PCI Industrial Computer Manufacturing Group
RACCA – Australia Air Conditioning Contractors Association
RSES – Refrigeration Service Engineers Society
SMWIA – Sheet Metal Workers Association
Solstice: Sustainable Energy and Development Online
UL – Underwriters Laboratories
United Service Alliance
University of Illinois – Dept.of Mech.Indust.Engr.
University of Kansas – Bldg.Mech.Sys
U.S. Doe
Water Technologies Assoc.
WBF – World Batch Forum
WTO – World Trade Organization

Year 2000 Problem
(Articles)
CIO Magazine's Feature Forum
Computerworld
Datamation
Legal Issues
Midrange Magazine
Ten Management and Legal Pitfalls
Third Party Time Bombs
Year 2000 does not compute
Year 2000 Information Page
Year 2000 or Bust
Year 2000 – Time's Running Out
(General Links)
Auto Parts & Accessories Association
Bank Administration Institute

Complete Y2K Website
Computing Services & Software Association
Data Dimensions
Director/Officer Liability
IBM-PCs and clones
IEE's "The Millenium Problem in Embedded Systems"
Information Management Forum
Information Technology Association of America's Year 2000
(Activities)
Information System Audit and Control Association
Information Week Discussions
Information Week Resource Center
Insurance and the Y2K Problem
International Standard ISO 8601
Legal Guidelines
Milliennium Bug
Milliennium Rollover
National Bulletin Board for the Year 2000
National Computing Centre (NCC)
National Retail Foundation's (NRF)
Society for Information Management
Texas A&M's Computing and Information Services
Tick, Tick, Tick quarterly newsletter
2K-Times
Washington D.C. Y2K Group
What Organization Can Do
Year 2000 computer problem
Year 2000 Info Center Forum
Year 2000 Information Line
Year 2000 Information Network
Year 2000 Problem
Year 2000 and 2-Digit Dates
Year 2000 Technical Audit Center
Y2Ki: Year 2000 Information Service
(*International Links*)
Australian Government
Australian Government & States
Australia's Year 2000 Activity Site
New Zealand
Norway (*in Norwegian*)
Province of British Columbia
Spain (*in Spanish*)
UK's Central Computer and Telecommunications Agency
UK's Computer Info Centre

(List of Links)
British Computing Society
Bill Cook's Y2K Resource Links
George Girod's bookmarks
NIST's Year 2000 Site
Randall de Weerd's Resources
The hourglass: European Information Exchange Structure page
The Y2K Computer Crisis
Y2K Links

Training
Factory Mutual
ISA's Training Institute
Jewel and Esk Valley College Course in PLCs

References

Chapter 1

"Electromagnetic Fields and the Risk of Cancer," Report of an Advisory Group on Non-Ionizing Radiation. National Radiological Protection Board, Vol. 1 No. 3, 1992.

Feychting, Maria and Anders Ahlbom, "Magnetic Fields and Cancer in Children Residing near Swedish High-voltage Power Lines," American Journal of Epidemiology, Vol. 7 No. 138, 1993, pp. 467-481.

Feychting, Maria and Anders Ahlbom, "Studies of Electromagnetic Fields and Cancer, How Inconsistent?" Environmental Science and Technology, Vol. 6 No. 10, 1993, pp. 1018-1020.

Gupton, Jr., Guy W., HVAC Controls: Operation and Maintenance, 2nd Edition, Fairmont Press, Inc.: Liburn, GA, 1996.

"Health Effects of Low-frequency Electric and Magnetic Fields," Prepared by an Oak Ridge Associated Universities Panel for the Committee on Interagency Radiation Research and Policy Coordination. ORAU, June 1992.

"Health Effects of Low-frequency Electric and Magnetic Fields," Summary of the ORAU Report with Commentary by David A. Savitz and Thomas S. Tenforde, Environmental Science and Technology, Vol. 1 No. 27, 1993, pp. 42-58.

Luepke, Gary, "Controlling Microbial Growth," Refrigeration Service and Contracting, April 1998 Vol. 66 No. 4, p. 24.

Trethewey, Richard, This Old House Heating, Ventilation and Air Conditioning, Little, Brown and Company: Boston, MA, 1994.

Internet: www.infoamp.netEathc/unico.html

"Air Conditioning for Homes Without Ductwork – Unico System History," Toronto, Canada: UnicoSystem, Dec. 16, 1998, pp. 1-2.

Internet: www.epa.gov

"Electricity and Magnetic Fields, 4" NJ: U.S. Environmental Protection Agency, Dec. 16, 1998, pp. 1-2.

Internet: www.appliance.com

"Geothermal Heating, Cooling and Water Heating Systems," Appliance Consumer Connection, Dec. 16, 1998, pp. 1-2.

Internet: www.ejnet.com

"What's so Special About 2 Inches?" Toronto, Canada: UnicoSystem, Dec. 16, 1998, pp. 1-2.

Chapter 2

Gupton, Jr., Guy W, HVAC Controls: Operation and Maintenance, 2nd Edition, Fairmont Press, Inc.: Liburn, GA, 1996.

Liptak, Bela A., Editor-in-Chief, Instrument Engineers' Handbook, 3rd Edition, Chilton Book Company: Radnor, PA, 1995.

Chapter 3

Liptak, Bela A., Editor-in-Chief, Instrument Engineers' Handbook, 3rd Edition, Chilton Book Company: Radnor, PA, 1995.

Skrokov, M. Robert, Editor, Mini- and Microcomputer Control in Industrial Processes, Van Nostrand: New York, 1980.

Internet: www.iac.honeywell,com

"PlantScape Application Note," Honeywell IAC, October 7, 1998, pp. 1-2.

Chapter 4

Horn, Delton T., Home Remote-Control and Automation Projects, 2nd Edition, Tab Books: Blue Ridge Summit, PA, 1991.

Liptak, Bela A., Editor-in-Chief, Instrument Engineers' Handbook, 3rd Edition, Chilton Book Company: Radnor, PA, 1995.

Chapter 5

Gutzwiller, F.W., General Electric Silicon-controlled Rectifier Manual, General Electric: Chicago, IL, 1988.

Horn, Delton T., Home Remote-Control and Automation Projects, 2nd Edition, Tab Books: Blue Ridge Summit, PA, 1991.

Liptak, Bela A., Editor-in-Chief, Instrument Engineers' Handbook, 3rd Edition, Chilton Book Company: Radnor, PA, 1995.

Motorola Semiconductor Circuits Manual, Motorola: Phoenix, AZ, 1964.

Chapter 6

Horn, Delton T., Home Remote-Control and Automation Projects, 2nd Edition, Tab Books: Blue Ridge Summit, PA, 1991.

Liptak, Bela A., Editor-in-Chief, Instrument Engineers' Handbook, 3rd Edition, Chilton Book Company: Radnor, PA, 1995.

Milestek Catalog 98-2, Milestek Inc.: Denton, TX, 1998.

Walczak, P.E., Thomas A. "The Pyramid Lives On," Industrial Computing, Vol. 17 No. 8, August 1998, pp. 18-22.

Internet: www.esteem.com

"Electronic Systems Technology, Inc.," July 2, 1998, p. 2.

Internet: www.breezecom.couk/Products/brztov.htmlanintro

"Introduction to LAN Topology," BreezeCOM, 11/30/98, pp. 1-5.

Chapter 7

Controller for Solar Heating-Design Package, National Aeronautics and Space Administration: Marshall Space Flight Center, AL, March 1979, p. I.

Dierauer, Peter P., "PLCs Giving Way to Smart Control," InTech, March 1998, Vol. 45 No. 3, pp. 40-44.

Foster, William M., Homeowner's Guide to Solar Heating and Cooling, Tab Books: Blue Ridge Summit, PA, 1976.

Hordeski, Michael F., Control System Interfaces: Design and

Implementation Using Personal Computers, Prentice-Hall, Inc.: Englewood Cliffs, NJ, 1992.

Klahn, Louis, "Practical PCs," Industrial Computing, Vol. 17 No. 6, June 1998.

Liptak, Bela A., Editor-in-Chief, Instrument Engineers' Handbook, 3rd Edition, Chilton Book Company: Radnor, PA, 1995.

Multichannel Temperature Controller for Solar Heating, National Aeronautics and Space Administration: Marshall Space Flight Center, AL, Oct. 1976, pp. I-II.

Solar Energy Heating Module Program, National Aeronautics and Space Administration: Marshall Space Flight Center, AL, Oct. 1976, p. 21.

Internet: www.barringtonsys.com/bacnet.htm

"BACnet," Barrington Systems, Nov. 11, 1998, p. 1.

Internet: www.odva.org/ABOUTDN/over6.htm

"CAN and DeviceNet," Open DeviceNet's Vendor Association, Nov. 18, 1998, pp. 1-2.

Internet: www.controlnet.org

"ControlNet Technical Specifications," ControlNet, November 18, 1998, pp. 1-2.

Internet: www.penton.com

Cosio, PE, Jeffrey, June 1998, pp. 1-8.

Internet: www.csv.warwick

"Introduction to DeviceNet," September 1998, pp. 1-3.

Internet: www.PLCOPEN.org

"Introduction to IEC 1131-3 and PLCopen," Sept. 9, 1998, p. 1.

Internet: www.isagra.com

"ISAaGRAF Overview on the Internet," Sept. 1998, pp. 1-6.

Internet: www.lonworks.echelon.com

"It Blows Hot and Cold in the Windy City," LonWorks, Nov. 30, 1998, pp. 1-3.

Internet: www.barringtonsys.com/lonworks.htm

"LonWorks," Barrington Systems, Nov. 11, 1998, p. 1.

Internet: www2.control.com

"Re: A Hunting Party," Nov. 11, 1998, p. 1.

Internet: www.controlnet.org

"Where Does ControlNet Fit?," ControlNet, November 18, 1998, p. 1.

Internet: www.controlnet.org

"Why ControlNet?," ControlNet, November 18, 1998, pp. 1-2.

Internet: www.wonderware.com

"Wonderware Products on the Internet," Wonderware Corp., Sept. 1998, pp. 1-7.

Chapter 8

Gupton, Jr., Guy W., HVAC Controls: Operation and Maintenance, 2nd Edition, Fairmont Press, Inc.: Liburn, GA, 1996.

Klahn, Louis, "Practical PCs," Industrial Computing, June 1998, Vol. 17 No. 6, pp, 61-64.

Liptak, Bela A., Editor-in-Chief, Instrument Engineers' Handbook, 3rd Edition, Chilton Book Company: Radnor, PA, 1995.

"SCADA Software Roundup," Industrial Computing, Oct. 1998, Vol. 17 No. 10, pp. 36-37.

Internet: www.siebe-env-controls.com

"Basics of PID Control," Siebe Environmental Controls, 1996-98.

Internet: www.iglou.com/seriplex

"Frequently Asked Questions: Seriplex Control Bus," Seriplex Inc., Nov. 30, 1998, pp. 1-7.

Internet: www.ibsclub.com

"Interbus-S Protocol Structure Data Sheet," Nov. 30, 1998, pp. 1-6.

Internet: www.profibus.com

"Profibus: Frequently Asked Questions," Nov. 18, 1998, pp. 2-9.

Internet: www.modicon.com

Swales, Andy, "Industrial Internets: Enabling Transparent Factories," Paper presented National Manufacturing Week, Chicago, IL, March 17, 1998, pp. 1-9.

Appendix

Strotham, James E. and Susan B. Colwell, Editors, "Editors' Picks: Best Control Web Sites," Control Platforms (Special Supplement to InTech), Oct. 1997, pp. 16-22.

Internet: www.NWlink.com

HVAC on the Web – Links, September 30, 1998, pp. 1-18.

A

ACH 8
AFUE, 6, 13
adiabatic, 103
air changes per hour, 8
airflow controller, 82
air volume regulator, 47
algorithms, 242
all-water systems, 22
Annual Fuel Utilization Efficiency, 6
aquastat, 10
automatic fuel-air ratio correction, 89
automatic mode reconfiguration, 126
automatic rebalancing, 62

B

backpressure regulator, 96
BAS, 254
baseboard radiators, 14
BASIC, 195-196
building automation system, 254
bus arbitrator, 247

C

Carrier Sense Multiple Access, 252
CDR, 263
childhood leukemia, 16
chilled water pump station, 121
chiller cycling, 128
closed-loop air control, 81
closed-loop control, 83, 89
CO analyzers, 88
comfort gap, 54
compensated control, 52
complementary data retransmission, 263
compressed fiberglass insulation, 11
computer-based energy management, 102
computer-optimized buildings, 39
condensing furnaces, 6
conduction, 1
constant volume regulator, 46
controller backup, 236
convection, 12
coprocessor, 227

Coriolis mass flowmeters, 80
counter cards, 205
CPU, 199, 209, 241

D
DDC, 90, 228, 233-234
DD/CV, 25
DDS, 210, 234
dead time, 76
derivative mode, 153
dew point thermostat, 133
differential pressure, 74
digital control, 233
direct control, 154
Direct Digital Control, 90, 228, 234
direct heat exchanger, 123
distributed control system, 234
distributed control systems, 210
DP, 74
droop, 229
dual-slope A/D converter, 192
dual thermostat, 135
dual-tone frequencies, 177
duct smoke detectors, 23
dynamic compensation, 70, 82

E
eddy-current coupling, 99
EMCS, 254
emissivity, 2
EMS, 254
Energy Management and Control Systems, 254
energy management systems, 254
enterprise network, 266
enthalpy logic, 57
Ethernet LAN, 185

F
facilities management system, 254
fan curve, 48
feedwater flow control, 74
fire codes, 42
fire condition, 40
flowchart-like languages, 205